12th International Conference on Nanomolecular Electronics (ICNME-2016)

Journal of Physics: Conference Series Volume 924

Kobe, Japan
14 – 16 December 2016

ISBN: 978-1-5108-5844-2
ISSN: 1742-6588

Printed from e-media with permission by:

Curran Associates, Inc.
57 Morehouse Lane
Red Hook, NY 12571

Some format issues inherent in the e-media version may also appear in this print version.

Printed by Curran Associates, Inc. (2017)

For permission requests, please contact the Institute of Physics
at the address below.

Institute of Physics
Dirac House, Temple Back
Bristol BS1 6BE UK

Phone: 44 1 17 929 7481
Fax: 44 1 17 920 0979

techtracking@iop.org

Additional copies of this publication are available from:

Curran Associates, Inc.
57 Morehouse Lane
Red Hook, NY 12571 USA
Phone: 845-758-0400
Fax: 845-758-2633
Email: curran@proceedings.com
Web: www.proceedings.com

Table of contents

Volume 924

12th International Conference on Nanomolecular Electronics (ICNME-2016)

14–16 December 2016, Kobe, Japan

Accepted papers received: 1 November 2017Published online: 5 December 2017

Preface

12th International Conference on Nanomolecular Electronics (ICNME-2016)

Peer review statement

Papers

Parametric Optimization of Dye-Sensitized Solar Cells Using Far red Sensitizing Dye with Cobalt Electrolyte

A Pradhan, M Saikiran, G Kapil, S S. Pandey and S Hayase.....1

Quantitative evaluation of spatial scale of carrier trapping at grain boundary by GHz-microwave dielectric loss spectroscopy

W. Choi, Y. Tsutsui, T. Miyakai, T. Sakurai and S. Seki.....10

Bulk-Heterojunction Thin-Film Solar Cells Utilizing Miscible Binary Donor Materials of Liquid Crystalline Phthalocyanine and its Analogue

Kento Fujita, Dai Nakagawa, Chika Nakano, Quang-Duy Dao, Akihiko Fujii and Masanori Ozaki.....18

Aggregation States of Organic Dye Molecules in Organic – Inorganic Hybrid Films Studied by Near-Field and Far-Field Fluorescence Spectroscopy

M. Ishihara.....25

Fabrication of Carbon Nanotube Polymer Actuator Using Nanofiber Sheet

Hayato Kato, Akikazu Shimizu, Taiga Sato and Masahito Kushida.....33

Photophysical Characterization and BSA Interaction of Direct Ring Carboxy Functionalized Symmetrical squaraine Dyes

Maryala Saikiran, Shyam S Pandey, Shuzi Hayase and Tamaki Kato.....45

Effect of High Pressure on Resistance in $Au(dmit)_2$ LB Films

Y F Miura, H Hasegawa, K Torizuka and Y Uwatoko.....55

Solution-processed organic field-effect transistors based on dinaphthothienothiophene precursor with chemically modified electrodes

Takashi Nagase, Souichiro Abe, Takashi Kobayashi, Yu Kimura, Azusa Hamaguchi, Yoshinori Ikeda and Hiroyoshi Naito.....61

Relation between active-layer thickness and power conversion efficiency in P3HT:PCBM inverted organic photovoltaics

S. Nakami, T. Narioka, T. Kobayashi, T. Nagase and H. Naito.....67

Functional design of electrolytic biosensor

D M Gamage Preethichandra, E M I Mala Ekanayake and M Onoda.....73

Calculation of Reflection Spectrum with Actual Layer Thickness Profile in Nacre of Akoya Pearl Oyster

R Ozaki, A Yoshimoto, G Watanabe, K Kadowaki and K Odawara.....85

Negative capacitance in an organic solar cell observed by displacement current measurement

Koji Shimizu, Yuya Tanaka, Yutaka Noguchi and Hisao Ishii.....92

Effect of immobilization technique on performance ZnO nanorods based enzymatic electrochemical glucose biosensor

Mayoorika Shukla, Pramila, I.A. Palani and Vipul Singh.....99

Casting Control of Floating-films into Ribbon-shape Structure by modified Dynamic FTM

A Tripathi, M Pandey, S Nagamatsu, S S Pandey, S Hayase and W Takashima.....109

Oriented antibody immobilization on self-assembled monolayers applied as impedance biosensors

Kaiki Tsugimura, Hitoshi Ohnuki, Haiyun Wu, Hideaki Endo, Daiju Tsuya and Mitsuru Izumi.....115

M intermediate accumulation analysis of bacteriorhodopsin reconstituted with three partial peptides

Y Tsujiuchi, H Masumoto and T Goto.....123

Vapor Deposition Polymerization and Molecular Orientation of Polyimide Thin Films

Takatoshi Yamazaki, Chanya Mahapun, Satoshi Usui, Kuniaki Tanaka and Hiroaki Usui.....135

IOP Conf. Series: Journal of Physics: Conf. Series **924** (2017) 011001 doi:10.1088/1742-6596/924/1/011001

PREFACE

12th International Conference on Nano-Molecular Electronics (ICNME2016) was held on December 14-16, 2016. This international conference is a series conference and has been held at Kobe, Japan, every two year. This conference has made a great contribution to the development of organic electronics materials & molecular electronics since its inception in 1994 (see special issues, **Japanese Journal of Applied Physics 34, No.7B, 1995, Thin Solid Films 331, 1998, Thin Solid Films 393, 2001, Thin Solid Films 438-439, 2003, Thin Solid Films 499, 2006, Thin Solid Films 516, Issue 9, 2008, Thin Solid Films 518, Issue 2, 2009, Physics Procedia 14, 2012, Thin Solid Films 554, Issue 3, 2014, Journal of Nanoscience and Nanotechnology 16, no.4, 2016**). This conference focuses mainly on the most attractive and interesting research subjects concerning the electrical and photonic properties and functions of organic materials. Main topics of the ICNME2016 were *Fundamentals of organic and molecular devices, Characterization and device physics, Energy harvesting materials and Devices, Molecular photonics and nanophotonics, Organic semiconductor devices and applications, Soft material physics and applications in electronics, and Bioelectronics and nanointerface fabrication.* Since its inception of this conference, ICNME has been paying attention to "interface issues" in molecular systems and in devices. In this ICNME2016, recent progress in the science and technology aspect on electronics, photonics and molecular electronics using organic molecules, polymers, and bio-related materials were intensively discussed in terms of "interfaces", from viewpoints of physics, chemistry and device applications. About 130 papers have been submitted from universities, research laboratories and industries from Japan and overseas, and these were presented as Plenary Talks, Invited Talks, and Oral and Poster presentations. Selective papers were accepted for publication after strict peer review process by experts in the world.

Conference chair,

Mitsumasa Iwamoto
Tokyo Institute of Technology

IOP Conf. Series: Journal of Physics: Conf. Series **924** (2017) 011002 doi:10.1088/1742-6596/924/1/011002

Peer review statement

All papers published in this volume of *Journal of Physics: Conference Series* have been peer reviewed through processes administered by the proceedings Editors. Reviews were conducted by expert referees to the professional and scientific standards expected of a proceedings journal published by IOP Publishing.

IOP Conf. Series: Journal of Physics: Conf. Series **924** (2017) 012001 doi:10.1088/1742-6596/924/1/012001

Parametric Optimization of Dye-Sensitized Solar Cells Using Far red Sensitizing Dye with Cobalt Electrolyte

A Pradhan*, M Saikiran, G Kapil, S S. Pandey* and S Hayase

Graduate School of LSSE, Kyushu Institute of Technology, 2-4, Hibikino, Wakamatsu, Kitakyushu 808-0196, Japan
E-mail: ansupradhan9@gmail.com;shyam@life.kyutech.ac.jp

Abstract. A far-red sensitizing dye **SQ-75** has been employed as a model sensitizer with $Co(bpy)^{2+/3+}$ redox electrolytes to fabricate dye-sensitized solar cells (DSSCs) and optimize the various device parameters which influence the overall photoconversion efficiency (PCE). It has been found that the optimization of the TiO_2 thickness, surface treatment with $TiCl_4$, and an optimum amount of the chenodeoxycholic acid (CDCA) as coadsorber are necessary to attain the overall improved PCE. $TiCl_4$ surface treatment on both FTO and TiO_2 has been found to outperform as compared to their untreated counterparts owing to the suppression of the charge recombination. DSSCs with an optimized TiO_2 thickness of 6 μm and CDCA concentration of 4 mM have exhibited best performance due to enhanced photon harvesting and reduced dye aggregation, respectively.

1. Introduction

Dye-sensitized solar cells (DSSCs) have been emerged as one of the potential candidates amongst next generation solar cells owing to their low production cost, ease of fabrication and widely tunable optical properties [1]. DSSC is basically a photoelectrochemical cell which is typically comprised of a sensitizing dye adsorbed to the surface of wide band gap semiconductor like TiO_2 as a photoanode, Platinum (Pt) coated conducting substrate as a photocathode and an electrolyte system containing redox couple [2-4]. The dyes which are heart of this kind of solar cells plays a crucial role in photon harvesting since they actually absorb the sun light and controls the overall obtainable photoconversion efficiency (PCE) and have shown nearly quantitative photon harvesting in the visible wavelength region [5-7]. In order to enhance the PCE even higher, there has been a continuous research in developing the sensitizing dyes having photon harvesting in the far-red to near infra-red (NIR) wavelength region a various dyes including dyes from squaraine family have attempted to extend the light absorption as well as photon harvesting in the red to NIR spectral region [8-15]. Recent past has witnessed that DSSCs have successfully achieved a PCE of 11.9% with iodine (I_3^-/ I^-) based electrolytes [15] even having efficient photon harvesting mainly in the visible region of the solar spectrum only. However, disadvantages such as potential loss, corrosion and competitive light absorption by iodine based redox electrolyte itself have limited the efficiency of DSSCs compelled the search for alternate redox electrolytes to circumvent these issues. In this context, new alternative redox mediators such as based on cobalt complexes redox shuttles have been successfully employed to achieve higher open circuit voltage (Voc) as well as high PCE [16-20]. Interestingly, a theoretical PCE reaching 20% with an assumption of minimal potential loss and better fill factor, can be attained by combining NIR and visible dye. Unfortunately there are no report regarding fabrication of DSSCs employing NIR dyes and cobalt complex based redox shuttles to be best of our knowledge. Therefore, we feel that squaraine dyes having sharp and intense light absorption mainly in the far-red to NIR region as a representative model of NIR dyes with Cobalt complex based electrolyte is an important and interesting area of research to be focused on and knowledge can be used for fabrication of high efficiency DSSCs.

Squaraine dyes exhibit narrow full width at half maximum and very high molar extinction coefficients and tunable light absorption form visible to IR wavelength region. They show strong dye aggregation behavior due to their relatively planar molecular structure and extended π–conjugation [10]. Therefore, we have accepted these challenges and tried to improve the efficiency of DSSCs based on squaraine dyes specially utilizing cobalt complex based redox electrolyte. A well-known dye de-aggregating agent chenodeoxycholic acid (CDCA) was co-adsorbed on the nanoporous TiO_2 surface along with the dye molecules to prevent dye aggregation [21]. In this work, we have employed an unsymmetrical squaraine dye (SQ-75) as a representative of NIR dye with $Co(bpy)^{2+/3+}$ redox electrolyte. Device fabrication parameters such as thickness of mesoporous TiO_2 layer, and surface treatment with $TiCl_4$ affects the performance of DSSCs [22-28], hence, similar kinds of experiments have also been carried out in current study on this novel dye-electrolyte combination in order to optimize the performance.

2. Materials and methods

Unsymmetrical squaraine dye SQ-75 and cobalt complex based redox electrolyte was synthesized following the methodology as per our earlier publications [9,29]. After the synthesis and purification by silica-gel column chromatography, structure of SQ-75 was confirmed by 1H nuclear magnetic resonance and fast ion bombardment-mass spectrometry. Electronic absorption spectra of the sensitizing squaraine dye (SQ-75) in ethanol solution as well as the dye adsorbed on thin films of the mesoporous TiO_2 was measured with a UV-visible-NIR spectrophotometer (JASCO, V670). DSSC fabrication was started with the cleaning of fluorine doped tin oxide (FTO) glasses by detergent solution, distilled water, acetone, isopropanol and acetone for 10 minutes each. After this, the photoanodes were prepared by screen-printing mesoporous DS/P and 30 NRD TiO_2 pastes (Solaronix) on FTO followed by baking at 500^0C for 30 min. This coating procedure was done one or several times to obtain TiO_2 thickness in the range of 2 to 18 μm especially for the thickness optimization of mesoporous TiO_2 layer. Some of the samples were pre and post-treated to freshly prepared TiO_2 films with aqueous $TiCl_4$ solution. This TiO2 surface treatment was conducted by dipping the substrates dipped in the 40 mM aqueous $TiCl_4$ solution at 80^0C for 30 min under continuous magnetic stirring at 500 rpm followed by the sintering at 500^0C for 30 minutes. Different photoanodes such as TiO_2 only (untreated samples), bottom $TiCl_4$ (pre-treated samples) on FTO, top $TiCl_4$ (post-treated samples) on mesoporous TiO2 and top/bottom $TiCl_4$ (pre and post-treated samples) were prepared for the current study. All of the samples were dipped in 0.2 mM dye solution in ethanol containing a fixed amount of CDCA varying from 2 mM to 30 mM for 4 hours at room temperature. For optimization of TiO_2 thickness and concentration of CDCA top/bottom $TiCl_4$ based photoanodes were taken. Counter electrode was prepared by FTO glass sputtered with 240 nm Ti followed 40 nm Pt and DSSC was assembled using this counter electrode with the photoanode and a 25 μm thick Himilan film (Mitsu-Dupont) spacer which melts at 110^0C by pressing the two parts for 20-30 seconds. Electrolyte was then injected and cell was finally sealed using UV resin to avoid any solvent leakage. The electrolyte was composed of 0.22 mM $Co(bpy)_3(PF_6)_2$, 0.033 mM $Co(bpy)_3(PF_6)_3$, 0.2 M tertiary butyl pyridine and 0.1 M $LiClO_4$ in Acetonitrile for all of the devices. Chemical structure of the unsymmetrical squaraine dye (SQ-75) and bipyridyl ligand bearing cobalt complex based redox electrolyte used in this work have been shown in the Fig.1

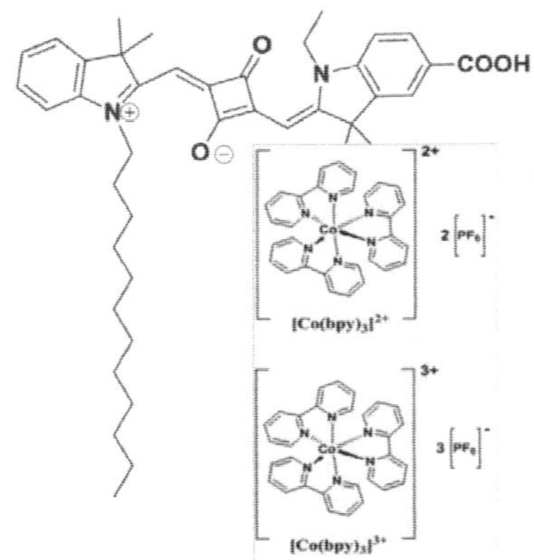

Figure 1. Molecular structure of sensitizing dye SQ-75 and cobalt complex based redox electrolyte.

Photovoltaic (PV) performances of the devices were measured with a solar simulator (CEP-2000, Bunko-Keiki, Japan) equipped with a xenon lamp (XLS-150A). The intensity of light irradiation was adjusted to AM 1.5 (100 mW/cm^2) using a spectroradiometer (LS-100, Eiko Seiki, Japan). The exposure power was also corrected with a standard amorphous Si photo detector (BS-520 S/N 007, Bunko-Keiki, Japan), which has similar light sensitivity to the DSSCs. Irradiation area of 0.2025 cm^2 was precisely controlled by a black metal mask for all of the samples during the measurement of photovoltaic characteristics. Incident photon to current conversion efficiency (IPCE) as a function of wavelength for the devices prepared were also measured with a constant photon flux of 1 x 10^{16} photon per cm^2 at each wavelength in the direct current mode using the action spectrum measurement system connected to the solar simulator.

3. Result and Discussion

3.1 Dye Design and Electronic Absorption Spectra

It has been found that molecular structure of dyes plays a crucial role to control the performance DSSCs especially using cobalt based redox electrolytes where dyes in general possess multiple and long alkyl chains [15, 20, 30]. This importance of alkyl chain for the sensitizers of DSSCs based on cobalt complex based electrolyte was further experimentally verified by the fact ruthenium based dye N-719 which works very well with iodine based electrolyte exhibit very poor performance with cobalt electrolyte. At the same time its long alkyl chain substituted analogue Z-907 works very well with cobalt based electrolytes [31]. Actually long alkyl chains of the dyes lead to effective surface passivation of the mesoporous TiO_2 surface and suppress the charge recombination by preventing obvious electrostatic attractive interaction between the positively charged cobalt complex ions with the negatively charged TiO_2 surface [32]. Keeping this mind we logically selected the unsymmetrical squaraine dye SQ-75 having very long dodecyl substituent at the opposite end of the dye from the –COOH functionalized indole ring as anchoring site. This not only supports the facile dye anchoring but also effectively passivates the negatively charged TiO_2 surface.

Owing to the flat molecular structure of squaraine dyes which are prone to the easy dye aggregation, CDCA has been most widely used in combination with dye as coadsorber to prevent the dye aggregation [21, 33]. Electronic absorption spectra of the SQ-75 in ethanol solution in the presence as well as well as absence of CDCA along with in solid-state thin film adsorbed on TiO_2 are shown in Fig. 2. Electronic absorption spectrum of SQ-75 in ethanol solution (5 mM with 4 mM CDCA) exhibits a very sharp absorption peak having absorption maximum (λ_{max}) at 638 nm with very high molar extinction coefficient of 3.18 x 10^5 dm^3.mol^{-1}.cm^{-1} along with vibronic shoulder at 592 nm. This is a typical characteristic of squaraine dyes and this sharp electronic absorption peak is associated with the

π-π^* electronic transition [4, 10]. Interestingly in the absence of CDCA also neither there is any shift in the spectral shape nor position of the λ_{max} which indicates that at in solution state at this concentration there is no dye aggregation and dye exists in the monomeric form. Upon absorption on the mesoporous TiO_2 not only there is spectral broadening as compared to that observed in solution but also there is enhancement in the intensity of the vibronic shoulder. This spectral broadening is attributed to the interaction between TiO_2 and dye molecules in the condensed state as reported previously [28]. The small difference in the intensity of the vibronic shoulder in the presence as well as absence of CDCA on TiO_2 suggests that long alkyl chain present in the dye are also involved in the suppression of dye aggregation in the condensed state.

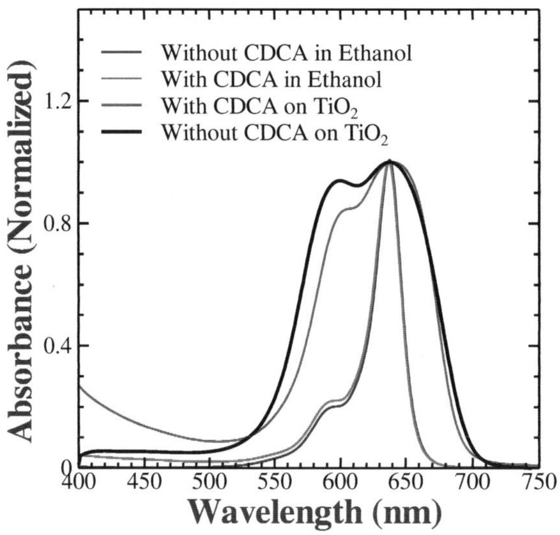

Figure 2. Electronic absorption spectra of SQ 75 ethanol solution and thin film adsorbed on mesoporous TiO_2 (3 μm).

3.2. Influence of $TiCl_4$ Surface Treatment

DSSCs have been fabricated in the device configuration of FTO/TiO_2/dye/Electrolyte/Pt with different interfaces between its all components and they are expected to work optimally for good device performances. Amongst these, FTO/TiO_2/dye interfaces play an important role in deciding the overall PCE [21-23]. Open circuit voltage (Voc) for a device varies inversely to the recombination rates at these interfaces, which in turn affects the short current density (Jsc) and fill factor (FF) also [24]. Control of this charge recombination by surface passivation using compact and conformal TiO_2 layer has been well documented. This surface passivation especially for DSSCs based on cobalt complex based electrolytes is more crucial owing to their relatively bulky nature as compared to iodine redox electrolyte counterparts and their relatively slow diffusion imparts more chances for the charge recombination [17]. Apart from passivation, the compact TiO_2 layer has been reported to play various roles such as to increase the TiO_2 surface area, to enhance electron transport, to promote light scattering and dye anchoring, which finally results in better device performance [22]. In this work, attempts have been made to investigate the role of this compact layer on overall device performance and to demonstrate that which one plays the predominant role. In order to predict this effect more explicitly, other factors such as thickness of mesoporous TiO_2 layer (6-7μm), dye (SQ-75), dye/CDCA ratio (1/50) and electrolyte $Co(bpy)^{2+/3+}$ were kept the same.

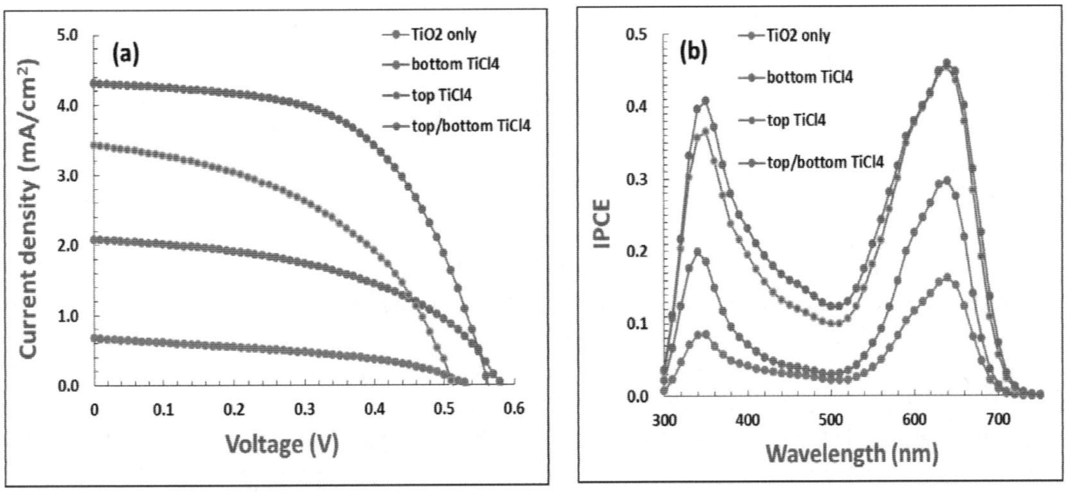

Figure 3. Photovoltaic characteristics of DSSCs under simulated solar irradiation (a) and photocurrent action spectra (b) after monochromatic incident light for DSSCs with different types of $TiCl_4$ surface treatments.

Figure 3(a) shows the current-voltage (I-V) characteristics of fabricated DSSCs under simulated solar condition of 1.5 AM with different surface treatment conditions. The short circuit current density (J_{sc}), open circuit voltage (V_{oc}), fill factor (FF) and the overall PCE (η) have been summarized in the Table 1. All of the DSSCs with $TiCl_4$ treatment exhibited better device performance as compared to that of untreated one due to reduced charge recombination. It can be clearly seen from this figure that surface treatment on mesoporous TiO_2 (top $TiCl_4$) has more pronounced effect than that on the FTO (bottom $TiCl_4$) which is concluded due to the rise in current density of 2.08 mA/cm^2 to 3.48 mA/cm^2 and hence the PCE of 0.57% to 0.81%. This increase in current density is well supported by the increased IPCE of 30% to 45 % between the wavelengths from 600-700 nm as shown in Fig 3(b). The beneficial effect of the $TiCl_4$ treatment is best presented by J_{sc} = 4.31 mA/cm^2, V_{oc} = 0.56 V, FF = 0.56 and η = 1.34%, when surface treatment was conducted on both of the FTO as well as mesoporous TiO_2 (top/bottom $TiCl_4$). IPCE shown in Fig. 3(b) also confirms the far-red photon harvesting by the sensitizer and maxima of the observed IPCE is in accordance with the Jsc measured in the I-V characteristics.

Table 1. Photovoltaic parameters for DSSC based on SQ-75 having $TiCl_4$ surface treatments conducted jon different surfaces of photoanode.

Treatment conditions	Jsc (mA/cm^2)	Voc (V)	FF	Efficiency (%)
No surface treatment	0.67	0.53	0.41	0.14
Treatment only on FTO	2.08	0.58	0.47	0.57
Treatment on TiO_2 only	3.43	0.51	0.46	0.81
Treatment on both TiO_2 and FTO both	4.31	0.56	0.56	1.34

3.3 Thickness of Mesoporous TiO_2 layer

Optimizing the thickness of light absorbing layer is an important issue to be solved as it affects the solar cell performance mainly the Jsc [26]. It can be easily understood that if the thickness of absorber layer is less, in that case most of the light transmits without proper absorption. Therefore, a minimum optimized thickness is must for proper device functioning. This

IOP Conf. Series: Journal of Physics: Conf. Series **924** (2017) 012001 doi:10.1088/1742-6596/924/1/012001

minimum thickness of the absorber layer will be less if the absorption coefficient of the absorbing material dyes is high. However, there is limitation on the maximum thickness of absorbing layer also, as with the increase in thickness after certain limit the chances of recombination increase because photoexcited electrons has to travel a long distance compared to their diffusion length [1]. This diffusion length varies for different dye and electrolyte systems used in DSSCs and need to be optimized for obtaining the better device functioning. In case of DSSCs, the absorbing layer thickness totally depends on mesoporous TiO_2 layer which acts as a scaffold for the dye monolayer. Hence, we can control and confirm the optimized thickness for better results by varying the thickness of mesoporous TiO_2 thickness.

In the present work, DSSCs were fabricated by varying the thickness of mesoporous TiO_2 layer in top/bottom $TiCl_4$ treated photoanodes keeping other variables such as dye concentration, electrolyte, CDCA concentration constant. Table 2 summarizes photovoltaic parameters obtained from I-V characteristics of the DSSCs. A perusal of table 2 corroborates the increase in thickness of the mesoporous TiO_2 layer from 2 μm to 4 μm there is increase in Jsc and PCE from 2.4 mA/cm^2 to 3.38 mA/cm^2 and 0.51% to 0.88%, respectively, due to enhanced photon harvesting. With further increase in the thickness of the mesoporous TiO_2 up to 6 μm, it was noticed that devices achieved their best Jsc of 5.33 mA/cm^2 and PCE of 1.43% whose photovoltaic characteristic is shown in the Fig. 4. When the thickness was increased more than 6 μm to 8 μm, there was a sharp decrease in the PCE as a function of thickness as shown in the inset of the Fig. 4. For Ruthenium complex based dyes (N-719) it has been demonstrated that 15 μm thick TiO_2 layer was optimum for achieving the best efficiency [26]. Therefore, we conclude that in our case, for the DSSC using sensitizing dye SQ-75 having a high molar extinction coefficient nearly 10 times as compared to typical ruthenium based sensitizers, a relatively thinner TiO_2 (6 μm) is capable of sufficient photons harvesting.

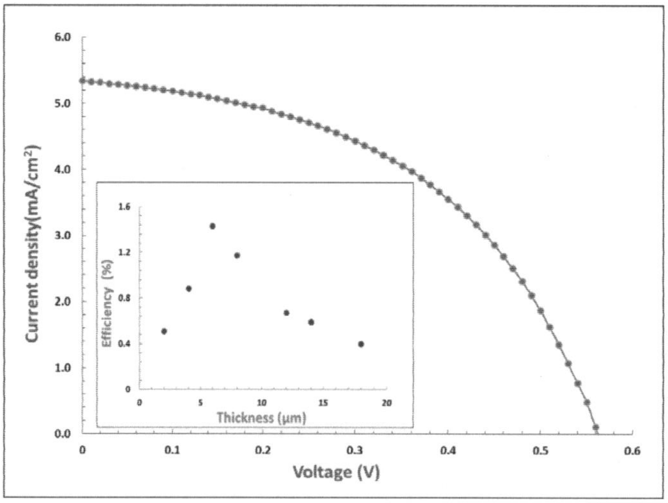

Figure 4. Photovoltaic characteristic of DSSC with TiO_2 layer thickness of 6 μm under simulated solar irradiation. Inset shows the thickness dependence of PCE.

Table 2. Photovoltaic characteristics of DSSC with varying thickness of mesoporous TiO_2.

Thickness	2 μm	4 μm	6 μm	8 μm	12 μm	14 μm	18 μm
Efficiency (%)	0.51	0.88	1.43	1.17	0.67	0.59	0.4
FF	0.37	0.47	0.48	0.55	0.47	0.5	0.48
V_{oc} (V)	0.56	0.55	0.56	0.56	0.55	0.55	0.52
J_{sc} (mA/cm^2)	2.4	3.378	5.33	3.75	2.59	2.1	1.62

3.4 Influence of CDCA Concentration

Squaraine dyes are prone to dye aggregate formation because of their planarity and extended π conjugation facilitated by the π-π stacking and lead to the hampered charge injection. Thus to prevent these dye aggregates, an adequate amount of CDCA has been most commonly used functions which not only functions as coadsorber but also suppresses the dye aggregation [21]. As CDCA plays the role of coadsorber without actually participating in photon harvesting, hence optimization for its adequate amount is necessary for the optimum device performance. For the optimization, top/bottom $TiCl_4$ based photoanode was implemented along with the same mesoporous TiO_2 thickness (6 μm), dye concentration and electrolyte. Figure 5 presents the plot of change in PCE for the DSSCs fabricated by changing the concentration of CDCA. Device containing 4 mM of CDCA in dye solution exhibited an improved PCE of 1.57% compared to the device without any CDCA (1.3%). Further increase in the CDCA concentration with respect to the dye resulted in to decrease of the PCE. This behavior could be understood considering the fact that presence of excess amount of CDCA inhibits the dye molecules present on the mesoporous TiO2 surface due to competitive adsorption and hampers the effective absorption of the incident photons. It is also interesting to note that SQ-75 dye structure is less prone to dye aggregation as devices without CDCA also showed the comparatively good PCE of 1.3% to that of the devices with CDCA as shown in Table 3.

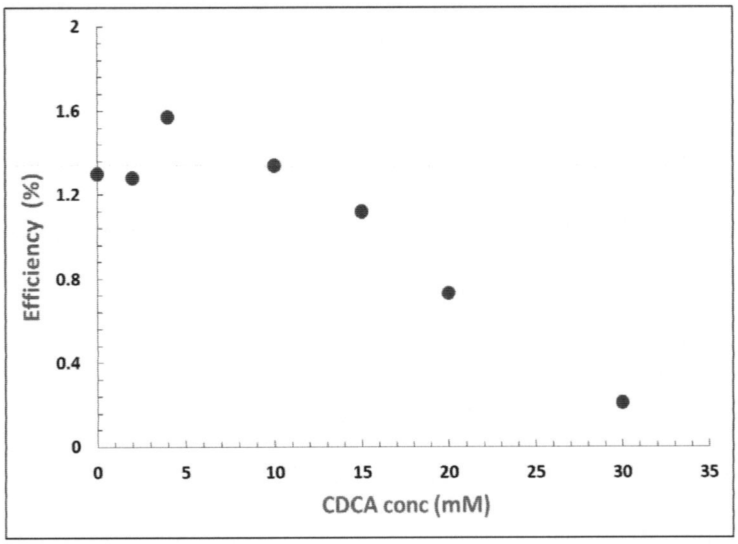

Figure 5. Effect of coadsorber CDCA concentration on the PCE of DSSCs based on SQ-75 and $Co(bpy)^{2+/3+}$ redox electrolyte.

*-***Table 3. Photovoltaic parameters for DSSCs with varying concentration of CDCA with SQ-75.

CDCA conc. (mM)	0	2	4	10	15	20	30
Efficiency (%)	1.3	1.28	1.57	1.34	1.12	0.73	0.21
Voc(V)	0.55	0.55	0.56	0.56	0.56	0.55	0.53
J_{sc} (mA/cm^2)	4.30	4.56	5.67	4.31	4.14	2.55	1.17
FF	0.55	0.51	0.49	0.56	0.48	0.49	0.35

4. Conclusions

In summary, optimization of DSSCs fabricated using a model unsymmetrical squaraine dye SQ-75 and $Co(bpy)^{2+/3+}$ redox electrolyte pertaining to the different device fabrication parameters have been amicably conducted in order to achieve the optimum PCE. $TiCl_4$ surface treatment was found to control the overall PCE and especially it was more effective when treatment was done on both of the FTO glass substrate and mesoporous TiO_2 layer leading to the best PCE of 1.34 %. It was also found that mesoporous TiO_2 thickness variation also affects the device performance and thickness of 6 μm was best suited for DSSCs based on this dye-electrolyte system. Finally, the PCE of SQ-75 dye based DSSCs seem to be less affected by the dye aggregation owing to its special molecular design as confirmed by photovoltaic measurements as well as electronic absorption spectral investigations.

Acknowledgments
One of the authors (SSP) would like express sincere thanks to Japan Society for the Promotion of Science (JSPS) for the financial support by Grant-in-Aid for Scientific Research (C) (Grant Number 26410206).

References:

[1] Grätzel M 2009 *Acc. Chem. Res.* **42** 1788
[2] Kakiage K, Aoyama Y, Yano T, Otsuka T, Kyomen T, Unno M and Hanaya M 2014 *Chem. Commun. 50 6379*
[3] Wang M, Grätzel C, Zakeeruddin S M, Grätzel M 2012 *Energy Environ. Sci.* **5** 9394
[4] T Morimoto, N Fujikawa, Y Ogomi, S S Pandey and S Hayase 2016 *Journal of Physics: Conference Series.* pp. 12002-12008.
[5] Ito S, Murakami T N, Comte P, Liska P, Grätzel C, Nazeeruddin M K, Grätzel M 2008 *Thin Solid Films* **516** 4613
[6] Gao F, Wang Y, Shi D, Zhang J, Wang M, Jing X, Humphry-Baker R, Wang P, Zakeeruddin S M, Grätzel M 2008 *J. Am. Chem. Soc.* **130** 10720
[7] Barbe C J, Arendse F, Comte P, Jirousek M, Lenzamann F, Shklover V, Grätzel M 1997 *J. Am. Ceram. Soc.* **80** 3157
[8] Sreejith S, Carol P, Chithra P, Ajayaghosh A 2008 *J. Mater. Chem.* **18** 264
[9] Pandey S S, Inoue T, Fujikawa N, Yamaguchi Y, Hayase S 2010 *Thin Solid Films.* **519** 1066
[10] Morimoto T, Fujikawa N, Ogomi Y, Pandey S S, Ma T, Hayase S 2016 *J. Nanosci. Nanotechnol.* (In press)
[11] J Chang, C P Lee, D Kumar, P W Chen, L Y Lin, K J Thomas and K C Ho 2013 *Journal of Power Sources.* **240** 779-785.
[12] Hara K, Tachibana Y, Ohga Y, Shinpo A, Suga S, Sayama K, Sugihara H, Arakawa H 2013 *Sol. Energy Mater. Sol. Cells* **77** 89
[13] C H Chang, Y C Chen, C Y Hsu, H H Chou and J T Lin 2012 *Organic letters.* **14** 4726-4729.
[14] F M Jradi, X Kang, D O'Neil, G Pajares, Y A Getmanenko, P Szymanski, T C Parker, M A El-Sayed and S R Marder 2015 *Chemistry of Materials.* **27** 2480-2487.
[15] Mathew S, Yella A, Gao P, Humphry-Baker R, Curchod BF, Ashari-Astani N, Tavernelli I, Rothlisberger U, Nazeeruddin M K, Gratzel M 2014 *Nat. Chem.* **6** 242
[16] Jihuai W, Zhang L, Jianming L, Miaoliang H, Yunfang H, Leqing F and Genggeng L 2015 *Chem.*

Rev. **115** 2136
[17] Zong X, liang M, Fan C, Tang K, Li G, Sun Z, Xue S 2012 *J. Phys. Chem. C.* **116** 11241
[18] M J DeVries, M J Pellin and J T Hupp 2010 *Langmuir.* **26** 9082-9087.
[19] M J Marchena, G de Miguel, B Cohen, J A Organero, S Pandey, S Hayase and A Douhal 2013 *The Journal of Physical Chemistry C.* **117** 11906-11919.
[20] S M Feldt, E A Gibson, E Gabrielsson, L Sun, G Boschloo and A Hagfeldt 2010 *Journal of the American Chemical Society.* **132** 16714-16724.
[21] J. H. Yum, S. J. Moon, R. H. Baker, P. Walter, T. Geiger, F. Nuesh, M. Grätzel, and M. K. Nazeeruddin, *Nanotechnology* 19, 424005(2008).
[22] Yoshida Y, Tokashiki S, Kubota K, Shiratuchi R, Yamaguchi Y, Kono M, Hayase S 2008 *Sol. Energy Mater. Sol. Cells* **92** 646
[23] Sommeling P M, O'Regan B C, Haswell R R, Smit H J P, Bakker N. J, Smits J J T, Kroon J M, van Roosmalen J A M 2006 *J. Phys. Chem.* B **110** 19191
[24] Park N G, Schlichthorl G, Lagemaat J V D, Cheong H M, Mascarenhas A, Frank A J 1999 *J. Phys. Chem. B* **103** 3308
[25] Zeng L Y, Dai S Y, Wang K J, Pan X, Shi C W, Guo L 2004 *Chin. Phys. Lett.* **21** 835
[26] Man Gu Kang, Kwang Sun Ryu, Soon Ho Chang, Nam Gyu Park, Jin Sup Hong, Kang-Jin Kim 2004 *Bull. Korean Chem. Soc.* **25** 742
[27] B A Gregg, F Pichot, S Ferrere and C L Fields 2001 *The Journal of Physical Chemistry B.* **105** 1422-1429.
[28] Yum J H, Walter P, Huber S, Rentsch D, Geiger T, Nuesch F, De Angelis F, Gratzel M, Nazeeruddin M K 2007 J. Am. Chem. Soc. **129** 10320.
[29] Marchena M J, de Miguel G, Cohen B, Angel Organero J, Pandey S, Hayase S, Douhal J 2013 Phys. Chem. C. **117** 11906–11919.
[30] Hanna E, Susanna K, Eriksson Sandra M, Feldt Erik Gabrielsson, Peter W, Lohse, Rebecka Lindblad, Licheng Sun, Håkan Rensmo, Gerrit Boschoo, and Anders Hagfeldt 2013 *J. Phys. Chem. C.* **117** (41) 21029–21036.
[31] Yeru L, James R J, Yao H, Qing W, Shaik M Z and Michael G 2011 *J. Phys. Chem. C.* **115** (38) 18847–18855.
[32] Kawano M, Nishiyama T, Ogomi Y, Pandey S S, Ma T and Hayase S 2015 RSC Advances, RSC Adv. 5 83725–83731.
[33] Wang Z S, Cui Y, Dan-Oh Y, Kasada C, Shinpo, Hara K 2007 J. Phys. Chem. C. **111** 7224–7230.

Quantitative evaluation of spatial scale of carrier trapping at grain boundary by GHz-microwave dielectric loss spectroscopy

W. Choi[1], Y. Tsutsui[1], T. Miyakai[2], T. Sakurai[1], S. Seki[1]

[1] Kyoto University, Kyoto, Japan.
[2] Osaka University, Osaka, Japan.

E-mail: seki@moleng.kyoto-u.ac.jp

Abstract. Charge carrier mobility is an important primary parameter for the electronic conductive materials, and the intrinsic limit of the mobility has been hardly access by conventional direct-current evaluation methods. In the present study, intra-grain hole mobility of pentacene thin films was estimated quantitatively using microwave-based dielectric loss spectroscopy (time-resolved microwave conductivity measurement) in alternating current mode of charge carrier local motion. Metal-insulator-semiconductor devices were prepared with different insulating polymers or substrate temperature upon vacuum deposition of the pentacene layer, which afforded totally four different grain-size conditions of pentacene layers. Under the condition where the local motion was determined by interfacial traps at the pentacene grain boundaries (grain-grain interfaces), the observed hole mobilities were plotted against the grain sizes, giving an excellent correlation fit successfully by a parabolic function representative of the boarder length. Consequently, the intra-grain mobility and trap-release time of holes were estimated as 15 $cm^2 V^{-1} s^{-1}$ and 9.4 ps.

1. Introduction

Organic field-effect transistors (OFETs) are of great interest in organic electronic devices due to their low-cost and light-weight characters attractive for practical applications [1–3]. Pentacene is a benchmark organic compound as a semiconducting layer in OFETs that has been studied over the last quarter century. The landscape of the overall charge transport in pentacene have been intensively investigated by field-induced electron spin resonance (FI-ESR) [4–6], electric field induced second harmonic generation (EFI-SHG) [7,8], and photoelectron spectroscopy [9–11], and successfully interpreted by the representative trap-release model of positive charge carriers. In contrast, the local scale carrier mobility as well as the interfacial trap density in pentacene were evaluated by field-induced time-resolved microwave conductivity (FI-TRMC) [12,13], leading to a characteristic upper limit of intra-grain in-plane hole mobility. It is known that the pentacene thin film shows diverse morphologies depending on the deposition condition and underlying material or substrate [14–17]. The grain size of pentacene has been revealed to show the positive correlation to their FET mobility, however few studies achieved the quantitative evaluation on the correlation between the morphology and charge carrier transport [18–20]. This is most likely due to the strong deterioration in the carrier mobility by the trap, since direct current method requires long-range translational motion of charge carriers detected as an electric current. In contrast to such direct current methods largely influenced by the carrier trapping, the charge transport property is spatially averaged in the FI-TRMC method, regardless of the carrier state—free or trapped. Here we report that this distinct feature of FI-TRMC enables to estimate the spatial and temporal scale of the grain-boundary trapping, which is discussed based on the quantitative evaluation of the correlation between the grain size and the local scale mobility.

2. Experimental

2.1. Fabrication of Metal-Insulator-Semiconductor (MIS) Devices.

The MIS devices used in this study was fabricated as follows; (1) A quartz substrate (4.9 × 50 mm^2, 1 mm thick) was treated with UV–O_3 prior to use; (2) Ti (adhesion layer, 5 nm) and Au (gate electrode, 30 nm) layers were successively deposited on the quartz substrate by DC sputtering and thermal vacuum deposition, respectively; (3) As the first insulating layer, a 300-nm-thick SiO_2 was deposited by RF sputtering; (4) As the second insulating layer, a 250-nm-thick poly(methyl methacrylate) (PMMA; Sigma-Aldrich) or 250-nm-thick polystyrene (PS; Sigma-Aldrich) layer was spincoated from its 3 wt% PS solution in toluene; otherwise, a 250-nm-thick layer of Cytop® (AGC Chemicals) was spincoated; (5) Thermal annealing was carried out for 1 h to remove residual solvent at 100 °C for PMMA and PS or at 180 °C for Cytop®; (6) a 75 nm thick pentacene layer, was thermally evaporated at a rate of 0.03 nm s^{-1} under 2 × 10^{-4} Pa, with controlling the temperature of the substrate (T_{sub}); (7) A 30-nm-thick top Au electrode was overcoated by vacuum deposition.

2.2. TRMC@Interfaces Measurements.

The microwave circuit was constructed using a waveguide system with homodyne setup, as described in our previous report [12,13]. The TRMC@*Interfaces* measurements were carried out with a microwave at 9 GHz from a signal generator (Rohde & Schwarz, SMF 100A). After the MIS device was loaded in the microwave cavity, a pulse gate bias voltage V_g(t) at an interval of 50 ms was applied based on a multifunction generator (NF Corporation, WF 1973). The current injected into the semiconductor–insulator interface ($I(t)$) was monitored using a digital phosphor oscilloscope (Tektronix, MDO 3022). Simultaneously, the reflected microwave ($P_r(t)$) is first amplified with a high gain FET amplifier (Gain 30 dB, 8 to 12 GHz, Ciao CA812-304), then evolved by a Schottky diode, and monitored by the oscilloscope. All experiments were conducted under ambient atmosphere at room temperature.

2.3. Characterization of Semiconducting Layers.

A 10-nm thick pentacene was thermally deposited on PMMA, PS, and Cytop® with the same condition to the device fabrication. The morphology of pentacene was characterized by a tapping-mode atomic force microscope (AFM: SII NanoTechnology, SPI-4000, Nanonavi II) using silicon cantilevers with a frequency of 150 kHz and spring constant of 9 N/m (OMCL-AC200TS-R3, OLYMPUS). The average grain radius is estimated from the topographic image in AFM (Figure 2).

3. Results and discussion

The schematic illustration of the measurement setup is shown in Figure 1. Electric field-induced hole carriers were injected at the pentacene–insulator interface in the MIS device with applied pulse gate biases, where these holes were oscillated by 9 GHz microwave.

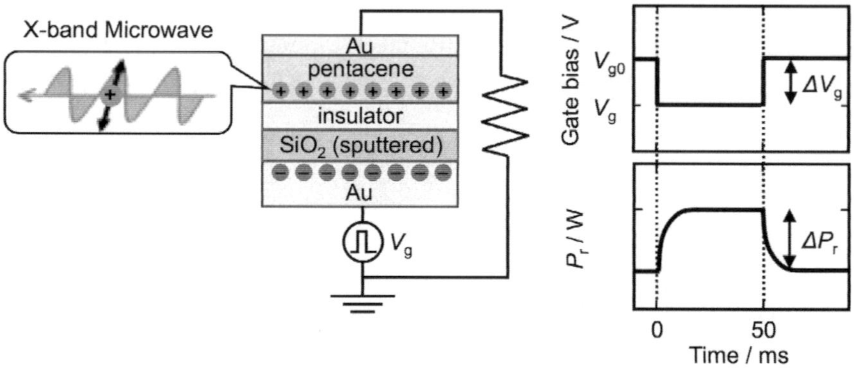

Figure 1. Schematic illustration of microwave dielectric loss spectroscopy of field-induced charge carriers (TRMC *@interfaces*).

As summarized in Table 1, the averaged grain radius and its distribution was 102.49 ± 0.26, 225.46 ± 0.09, 121.78 ± 0.35, and 76.86 ± 0.18 nm for PMMA, PMMA (T_{sub} = 80 °C), PS, and Cytop®, respectively. It was obvious that grain sizes were different among the four devices, which was caused by the different surface energy of the insulators.

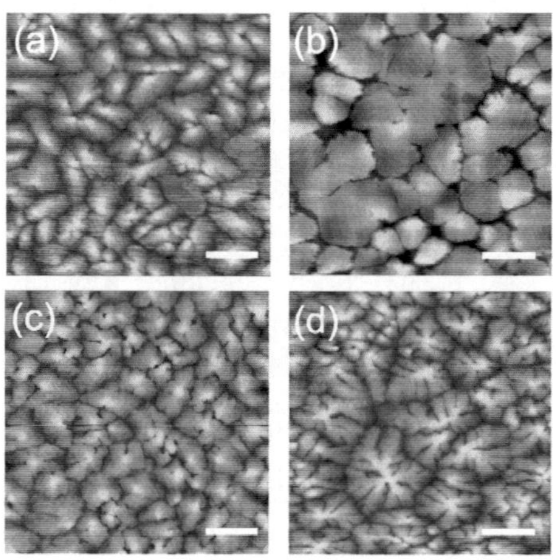

Figure 2. AFM images of pentacene layers on (a) PMMA, (b) PMMA (T_{sub} = 80 °C), (c) polystyrene, and (d) Cytop®. Scale bars indicate 1 μm.

With FI-TRMC, a dielectric loss is induced by the charge carriers injected into the pentacene–insulator interface of the MIS device in the microwave cavity [21–24]. The change in reflected microwave power (ΔP_r) is associated with the change of pseudo conductivity ($\Delta N\mu$) as follows:

$$\Delta P_r = K\Delta N\mu \,, \qquad (1)$$

where K, N and μ is the experimentally determined sensitivity factor which was previously calibrated to be 2.4 × 10^{-23} W V s cm^{-2} [12], the number of charge carriers, and the charge carrier mobility, respectively. In this study, the values of $\Delta N\mu$ for various V_g are determined from the observed ΔP_r. On the other hand, the number of injected charge carriers (N_{inj}) is estimated independently from the integration of $I(t)$ measured with the external circuit:

$$N_{inj} = \frac{1}{e}\int I(t)dt \,, \qquad (2)$$

IOP Conf. Series: Journal of Physics: Conf. Series **924** (2017) 012002 doi:10.1088/1742-6596/924/1/012002

where e denotes the elementary charge. N_{inj}–$\Delta N\mu$ plots for different second insulating layer are shown in Figure 3. The slop and intercept of the plot in high V_g region signify the charge carrier mobility and the number of trap sites, respectively [13,24].

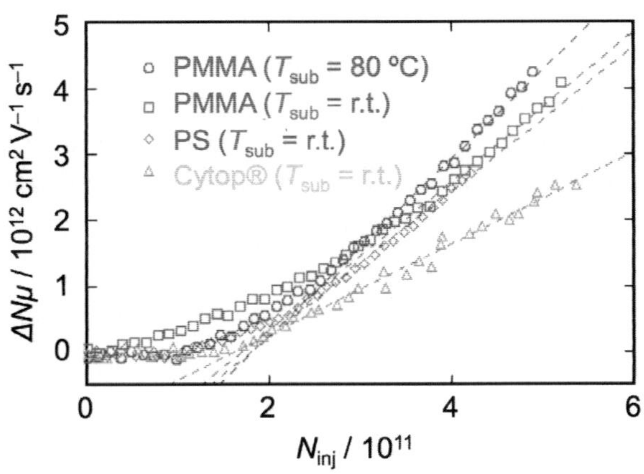

Figure 3. Correlation between the number of injected charge carriers (N_{inj}) and pseudo electrical conductivity $\Delta N\mu$ for various insulators.

Table 1. Summary of grain sizes, hole mobility, and trap density.

	Grain radius / nm	Hole mobility / cm^2 V^{-1} s^{-1}	Trap density / 10^{12} cm^{-2}
PMMA	102.49 ± 0.26	11	0.94
PMMA (T_{sub} = 80 °C)	225.46 ± 0.09	13	1.0
Polystyrene	121.78 ± 0.35	11	1.0
Cytop®	76.86 ± 0.18	6.9	0.89

The trap sites can be mainly categorized into 3 causes: (1) chemical impurities in the pentacene layer [25], (2) physical defects in the pentacene layer (mainly accumulated at the grain boundary [26], but structural imperfection and surface roughness [27] are also included), and (3) random potential from the dielectric layer [28]. Herein, (1) is considered in the same level for all samples. Since the charge carriers are only locally perturbed in FI-TRMC measurement, the measured mobility is insensitive to (3) the surface roughness itself and the random potential from the dielectric layer. As shown in Table 1, the density of interfacial trap sites is almost constant despite of the change of interface, indicating that this deep trap sites originate rather from the pentacene such as pentacenequinone chemical impurities [25] than the boundary structure of gate dielectrics [8]. Microwave-perturbation silent characteristics imply the deep-enough stabilization of the charge carriers at trap sites where the thermal activation energy cannot cause even trap-release processes, and thus no equilibrium can be postulated to lead certain amount of free (mobile) charge carriers. Consequently, both the relaxation time of the momentum and the effective mass of the charge carriers in intra-grain region of pentacene is independent to the organic insulating layer, suggesting that the observed difference in the charge carrier mobility resulted from (2), the difference in grain sizes.

On the assumption of the "sallow" traps accumulated at the grain boundary, we built a simple geometrical model with two concentric circles to assess quantitatively the reduced mobility of holes via thermally activated trap-release processes with an equilibria. Here, the outer circle represents a single grain while the inner one indicates an area with intra-grain hole mobility. In the area between the two circles, holes were trapped at the grain boundary. This theoretical model relies on the fact that spatial information of mobility is uniformly averaged in FI-TRMC measurement. Base on this model, the evaluated mobility can be expressed as following equation:

$$\mu = \mu_{\text{intra}} \frac{(r-L)^2}{r^2}, \qquad (3)$$

where μ_{intra}, r and L denote the intra-grain mobility, average radius of the grain, and spatial length dominated by the grain boundary (microwave-silent region), respectively. Here, the charge carrier mobility in grain boundary is assumed as 0 because it is much lower than the intra-grain mobility [26,29]. As shown in Figure 4, r dependence of μr^2 is clearly fitted with the parabolic function.

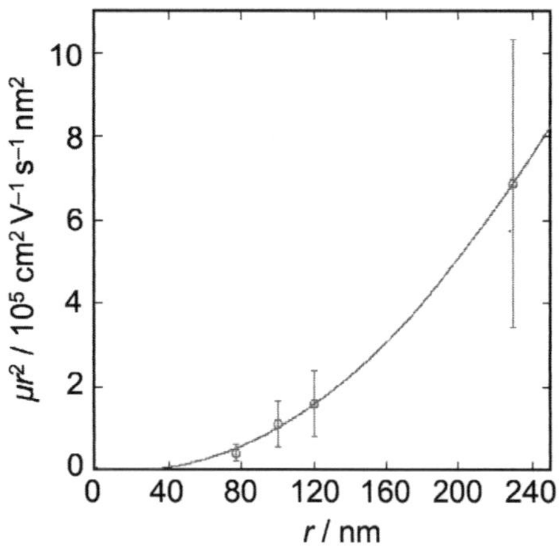

Figure 4. r dependence of μr^2. Parabolic dependence is shown as a solid line.

From the parabolic fit, μ_{intra} and L are derived as 15 $cm^2 V^{-1} s^{-1}$ and 19 nm, respectively. The turn-over period of the electric field of the 9 GHz microwave is much longer than the time scale of the charge carrier scattering. Consequently, the obtained characteristics from FI-TRMC reflect the equilibrium state of the charge carriers. Since the equilibrium between the free carriers and the traps of the grain boundary is established, the average time of the carrier trapping is same with the average time of the trap-release. That implies that the diffusion length for the trap-release time is the spatial length of the carrier-trapping region. The trap-release time (τ) in the grain boundary could be evaluated from the following diffusion equation:

$$L = \left(\frac{\mu_{\text{intra}} kT}{e} \tau \right)^{1/2}, \qquad (4)$$

where k and T denote the Boltzmann constant and temperature, respectively. From this equation, τ is derived as 9.4 ps. In relation to this analysis, there has been little work to quantify the attempt-to-escape frequency of the trapped carriers in organic semiconductors. Several studies to evaluate the attempt-to-escape frequency show very wide-range values (10^8–10^{14} Hz) [30–33]. The estimated trap-release time indicates that the attempt-to-escape frequency in the grain boundary of pentacene (1.0×10^{11} Hz) is comparable to that of the inorganic semiconductors (10^{11}–10^{13} Hz). The other detailed energetic information on the trap sites such as trap depth and energetic distributions will be revealed precise analysis of frequency dependent microwave conductivity based on the perturbation model [34] with the disclosed value of τ in the present study, which will interpret the total landscape of the hole transport mechanism in pentacene.

4. Conclusions

Dielectric loss spectroscopy using 9 GHz microwave was demonstrated for pentacene-based metal-insulator-semiconductor (MIS) devices with field-induced hole carriers. Four MIS devices with different crystalline grain sizes of pentacene were prepared by controlling the substrate temperature at r.t. or 80 °C and choosing different insulating polymers including poly(methylmethacrylate), polystyrene, and Cytop®. Considering that interfacial traps originate from the pentacene grain boundaries, the correlation between grain sizes and observed hole mobilities were plotted and successfully fit by a parabolic function. Consequently, the intra-grain hole mobility and trap-release time of the trapped carriers were estimated as 15 $cm^2 V^{-1} s^{-1}$ and 9.4 ps. The present work provided a general possibility that the microwave-based technique enables to estimate the intra-grain intrinsic mobility and trap-release time of charge carriers in semiconductors from the simple theoretical model, which is important for further development of organic electronics.

Acknowledgements

This work was partially supported by KAKENHI program under the Grant Nos. 26620104 and 261020110 from Japan Society for the Promotion of Science (JSPS) and the Ministry of Education, Culture, Sports, Science and Technology, Casio Science Promotion Foundation, Asahi Glass Foundation, and Kato Foundation for Promotion of Science. W.C. and Y.T. thank the JSPS for a Young Scientist Fellowship.

References

[1] Gelinck G H, Huitema H E A, Veenendaal E V, Cantatore E, Schrijnemakers L, van der Putten J B P H, Geuns T C T, Beenhakkers M, Giesbers J B, Huisman B H, Meijer E J, Benito E M, Touwslager F J, Marsman A W, van Rens B J E and de Leeuw D M 2004 Flexible active-matrix displays and shift registers based on solution-processed organic transistors *Nat. Mater.* **3** 106
[2] Sekitani T and Someya T 2010 Stretchable, large-area organic electronics *Adv. Mater.* **22** 2228
[3] Capelli R, Toffanin S, Generali G, Usta H, Facchetti A and Muccini M 2010 Organic light-emitting transistors with an efficiency that outperforms the equivalent light-emitting diodes *Nat. Mater.* **9** 496
[4] Marumoto K, Kuroda S, Takenobu T and Iwasa Y 2006 Spatial extent of wave functions of gate-induced hole carriers in pentacene field-effect devices as investigated by electron spin resonance *Phys. Rev. Lett.* **97** 256603
[5] Tanaka H, Hirate M, Watanabe S, Kaneko K, Marumoto K, Takenobu T, Iwasa Y and Kuroda S 2013 Electron spin resonance observation of charge carrier concentration in organic field-effect transistors during device operation *Phys. Rev. B* **87** 045309
[6] Marumoto K, Fujimori T, Ito M and Mori T 2012 Charge formation in pentacene layers during solar-cell fabrication: direct observation by electron spin resonance *Adv. Energy Mater.* **2** 591
[7] Manaka T, Lim E, Tamura R and Iwamoto M 2007 Direct imaging of carrier motion in organic transistors by optical secondharmonic generation *Nat. Photonics* **1** 581
[8] Manaka T, Liu F, Weis M and Iwamoto M 2010 Influence of traps on transient electric field and mobility evaluation in organic field-effect transistors *J. Appl. Phys.* **107** 043712
[9] Kanai K, Honda M, Ishii H, Ouchi Y and Seki K 2012 Interface electronic structure between organic semiconductor film and electrode metal probed by photoelectron yield spectroscopy

Org. Electron. **13** 309

[10] Nakayama Y, Uragami Y, Yamamoto M, Yonezawa K, Mase K, Kera S, Ishii H and Ueno N 2016 High-resolution core-level photoemission measurements on the pentacene single crystal surface assisted by photoconduction *J. Phys.; Condens. Matter* **28** 094001

[11] Yang Q, Muntwiler M and Zhu X Y 2009 Charge transfer excitons and image potential states on organic semiconductor surfaces *Phys. Rev. B* **80** 115214

[12] Honsho Y, Miyakai T, Sakurai T, Saeki A and Seki S 2013 Evaluation of intrinsic charge carrier transport at insulator-semiconductor interfaces probed by a non-contact microwave-based technique *Sci. Rep.* **3** 3182

[13] Choi W, Miyakai T, Sakurai T, Saeki A, Yokoyama M and Seki S 2014 Non-contact, non-destructive, quantitative probing of interfacial trap sites for charge carrier transport at semiconductor-insulator boundary *Appl. Phys. Lett.* **105** 033302

[14] Dimitrakopoulos C D and Malenfant P R L 2002 Organic thin film transistors for large area electronics *Adv. Mater.* **14** 99

[15] Shtein M, Mapel J, Benziger J B and Forrest S R 2002 Effects of film morphology and gate dielectric surface preparation on the electrical characteristics of organic-vapor-phase-deposited pentacene thin-film transistors *Appl. Phys. Lett.* **81** 268

[16] Knipp D, Street R A, Völkel A and Ho J 2003 Pentacene thin film transistors on inorganic dielectrics: Morphology, structural properties, and electronic transport *J. Appl. Phys.* **93** 347

[17] Ruiz R, Choudhary D, Nickel B, Toccoli T, Chang K C, Mayer A C, Clancy P, Blakely J M, Headrick R L, Iannotta S and Malliaras G G 2004 Pentacene thin film growth *Chem. Mater.* **16** 4497

[18] Kim D, Lee H, Yang H, Yang L and Cho K 2008 Tunable crystal nanostructures of pentacene thin films on gate dielectrics possessing surface-order control *Adv. Funct. Mater.* **18** 1363

[19] Hutchins D O, Weidner T, Baio J, Polishak B, Acton O, Cernetic N, Ma H and Jen A K Y 2013 Effects of self-assembled monolayer structural order, surface homogeneity and surface energy on pentacene morphology and thin film transistor device performance *J. Mater. Chem. C* **1** 101

[20] Liu C F, Lin Y, Lai W Y and Huang W 2017 Understanding the dependence of performance on the dielectric-semiconductor interface in pentacene-based organic field-effect transistors *Mater. Lett.* **189**, 286

[21] Tsutsui Y, Sakurai T, Minami S, Hirano K, Satoh T, Matsuda W, Kato K, Takata M, Miura M and Seki S 2015 Evaluation of intrinsic charge carrier transporting properties of linear- and bent-shaped π-extended benzo-fused thieno[3,2-b]thiophenes *Phys. Chem. Chem. Phys.* **17** 9624

[22] Sakurai T, Tsutsui Y, Choi W and Seki S 2015 Intrinsic charge carrier mobilities at insulator–semiconductor interfaces probed by microwave-based techniques: studies with liquid crystalline organic semiconductors *Chem. Lett.* **44** 1401

[23] Tsutsui Y, Schweicher G, Chattopadhyay B, Sakurai T, Arlin J B, Ruzié C, Aliev A, Ciesielski A, Colella S, Kennedy A R, Lemaur V, Olivier Y, Hadji R, Sanguinet L, Castet F, Beljonne D, Cornil J, Samorì P, Seki S and Geerts Y H 2016 Unraveling unprecedented charge carrier mobility through structure property relationship of four isomers of didodecyl[1]benzothieno[3,2-b][1]benzothiophene *Adv. Mater.* **28** 7106

[24] Inoue J, Tsutsui Y, Choi W, Kubota K, Sakurai T and Seki S 2017 Rapid evaluation of electron mobilities at semiconductor–insulator interfaces in an ambient atmosphere by a contactless microwave-based technique *ACS Omega* **2** 164

[25] Jurchescu O D, Baas J and Palstra T T M 2004 Effect of impurities on the mobility of single crystal pentacene *Appl. Phys. Lett.* **84** 3061

[26] Schön J H and Kloc C 2001 Charge transport through a single tetracene grain boundary *Appl. Phys. Lett.* **78** 3821

[27] Knipp D, Street R A and Völkel A R 2003 Morphology and electronic transport of polycrystalline pentacene thin-film transistors *Appl. Phys. Lett.* **82** 3907

[28] Jurchescu O D, Popinciuc M, van Wees B J and Palstra T T M 2007 Interface-Controlled, High-

Mobility Organic Transistors *Adv. Mater.* **19** 688
[29] Minari T, Nemoto T and Isoda S 2004 Fabrication and characterization of single-grain organic field-effect transistor of pentacene *J. Appl. Phys.* **96** 769
[30] Craciun N I, Wildeman J and Blom PW M 2008 Universal arrhenius temperature activated charge transport in diodes from disordered organic semiconductors *Phys. Rev. Lett.* **100** 056601
[31] Okachi T, Nagase T, Kobayashi T and Naito H 2009 Determination of localized-state distributions in organic light-emitting diodes by impedance spectroscopy *Appl. Phys. Lett.* **94** 043301
[32] Lee K, Weis M, Manaka T and Iwamoto M 2012 Multiple-trapping in pentacene field-effect transistors with a nanoparticles self-assembled monolayer *AIP advances* **2** 022128
[33] Carr J A, Elshobaki M and Chaudhary S 2015 Deep defects and the attempt to escape frequency in organic photovoltaic materials *Appl. Phys. Lett.* **107** 203302
[34] Choi W, Tsutsui Y, Sakurai T and Seki S 2017 Complex permittivity analysis revisited: Microwave spectroscopy of organic semiconductors with resonant cavity *Appl. Phys. Lett.* **110** 153303

IOP Conf. Series: Journal of Physics: Conf. Series **924** (2017) 012003 doi:10.1088/1742-6596/924/1/012003

Bulk-Heterojunction Thin-Film Solar Cells Utilizing Miscible Binary Donor Materials of Liquid Crystalline Phthalocyanine and its Analogue

Kento Fujita, Dai Nakagawa, Chika Nakano, Quang-Duy Dao, Akihiko Fujii, and Masanori Ozaki

Division of Electrical, Electronic and Information Engineering, Graduate School of Engineering, Osaka University, Suita, Osaka 565-0871, Japan
E-mail: afujii@eei.eng.osaka-u.ac.jp

Abstract. Small-molecule bulk-heterojunction solar cells utilizing a binary-blended donor material composed of a liquid crystalline phthalocyanine and its analogue have been studied. The short-circuit current density (J_{sc}) and power conversion efficiency (PCE) of the solar cells utilizing the blended donor material with an optimized blend ratio reached 11.6 mA/cm^2 and 4.8%, respectively, which were superior to those of organic solar cells utilizing each donor material. The improvement of J_{sc} and PCE has been discussed from the viewpoints of the crystal structure, optical and carrier transport properties of the blended donor material.

1. Introduction

Recently, organic solar cells (OSC) based on solution-processed small-molecule (SM) semiconductors have attracted attention as potential alternatives to widely used conjugated polymer-based OSC [1-5]. For commercial applications, solution-processed SM semiconductors offer relatively simple synthesis and purification, monodispersity, well-defined structures without end group contaminants, relatively high charge carrier mobility, and better batch-to-batch reproducibility [3].

Some kinds of liquid crystalline (LC) materials have been demonstrated as novel SM semiconductors for organic thin-film transistors [5-7], organic light-emitting diodes [5,8,9], and solution-processed bulk-heterojunction (BHJ) OSC [9-12]. LC semiconductors exhibit the appropriate characteristics for thin-film optoelectronic devices, such as high solubility in common organic solvents and strong self-organizing nature, which lead to relatively easy fabrication of thin-films with large monodomain and high charge carrier mobility.

One of discotic LC semiconducting materials, 1,4,8,11,15,18,22,25-octahexylphthalocyanine ($C6PcH_2$) exhibits an ambipolar high carrier drift mobility exceeding 1 cm^2/Vs in the crystalline phase, which was evaluated by time-of-flight (TOF) measurement [7]. BHJ OSC utilizing $C6PcH_2$ mixed with 1-(3-methoxycarbonyl)-propyl-1-phenyl-(6,6)C61 (PCBM) demonstrated high power conversion efficiency (PCE) exceeding 4% [10,13]. As further studies, fundamentals of an analogue molecule of $C6PcH_2$, 1,4,8,11,15,18,22,25-octahexyltetrabenzotriazaporphyrin ($C6TBTAPH_2$), have been reported, and high PCE exceeding 4.9% has been demonstrated in the BHJ OSC utilizing $C6TBTAPH_2$ mixed with 1-(3-methoxycarbonyl)propyl-1-phenyl-(6,6)C_{71} ([70]PCBM) [14, 15].

In order to improve the photovoltaic properties of OSC with LC semiconducting materials, we focused on the miscibility. Miscibility is defined as mixing at the molecular level to produce one

IOP Conf. Series: Journal of Physics: Conf. Series **924** (2017) 012003 doi:10.1088/1742-6596/924/1/012003

homogeneous blend [16]. Introducing the concept of miscibility, improvement of the charge carrier mobility has been reported [17]. On the other hand, molecular alignment control of mixed materials has been reported [18]. Further studies have reported that the PCE of OSC was improved by utilizing the mixed donor materials, such as $C6PcH_2$ and its homologue molecule, 1,4,8,11,15,18,22,25-octapentylphthalocyanine ($C5PcH_2$) [16].

In our recent studies, it was clarified that $C6TBTAPH_2$ and $C6PcH_2$ possess the miscibility in their mixed state, that is, each molecule is mixed at the molecular level [23]. Though the PCE of BHJ OSC is expected to be improved by utilizing $C6TBTAPH_2$ and $C6PcH_2$, the detailed studies remain to be carried out.

In this study, the BHJ OSC utilizing the blended donor materials of $C6TBTAPH_2$ and $C6PcH_2$ are studied. The photovoltaic properties of the solar cells utilizing such binary donor materials are discussed by considering the absorption spectrum, crystal structure in the thin film and hole mobility in the crystal phase.

2. Experimental

$C6TBTAPH_2$ and $C6PcH_2$ were synthesized and purified as reported previously [7, 20]. [70]PCBM (Frontier carbon Ltd.) was used without further purification. The molecular structures of $C6TBTAPH_2$, $C6PcH_2$, and [70]PCBM are shown in Fig. 1.

The solar cells, the schematic device structure of which is also shown in Fig. 1, were fabricated by the following process. Indium tin oxide (ITO)-coated glass substrates were cleaned with detergent, then ultrasonicated in water, chloroform, acetone, and isopropyl alcohol, and subsequently treated with UV-induced ozone. 6-nm-thick MoO_3 film was thermally evaporated onto the ITO-coated glass substrates at a rate of 0.1 Å/s under a vacuum of about 2×10^{-5} Pa. After transferring the substrates into a N_2-filled glove box, a solution, which was prepared by dissolving the donor materials, $C6TBTAPH_2$ and $C6PcH_2$, and acceptor material, [70]PCBM in chloroform with 0.2% v/v 1,8-diiodooctane (DIO), was spin-coated onto the MoO_3 layers. The blend ratio of donor and acceptor materials and total concentration were fixed to be 13:10 mol% and 21.5 g/L, respectively. The blend ratios of $C6PcH_2$ in the donor materials were mainly 0, 25, 50, 75, and 100 mol%. The thickness of the active layer with the BHJ of $C6TBTAPH_2$:$C6PcH_2$:[70]PCBM, which was estimated using atomic force microscopy (AFM, Keyence VN-8000), was approximately 120 nm. Finally, 80-nm-thick aluminum electrode was deposited through a shadow mask by thermal evaporation under a vacuum of about 2×10^{-5} Pa. The active area of the device was 4 mm^2.

The absorption spectrum of a thin film prepared on quartz substrate was measured using a spectrophotometer (Shimadzu UV-3150) at room temperature. The energy level at the highest occupied molecular orbital (HOMO) of a material in a thin film prepared on ITO-coated glass substrate was determined by using a measurement system of photoelectron yield spectroscopy (Sumitomo Heavy Industries PYS-202-H) under a vacuum of 10^{-3} Pa at room temperature.

The current density–voltage (*J–V*) characteristics under an illumination from a full spectrum solar simulator (Bunkoukeiki, OTENTO-SUNⅢP2) at an irradiation intensity of 100 mW/cm^2 (AM1.5G) were measured in vacuum at room temperature with a source measure unit (Keithley 2400). The external quantum efficiency (EQE) spectra were measured by using a photosensitivity measurement system (Bunkoukeiki, SM-250NA) with a Xe lamp light passing through a monochromator as a light source, the intensity of which was calibrated using a Si reference cell. The integrated EQE values indicated good agreements with the measured short-circuit current density (J_{sc}).

The charge carrier mobility under a low vacuum (10^{-1} Pa) was determined by the TOF technique. A sandwich cell consisting of two ITO-coated glass substrates and a 12-μm-thick PET spacer was fabricated. The actual cell gap was evaluated by an interference technique of light transmittance. Subsequently, binary-blended materials of $C6TBTAPH_2$ and $C6PcH_2$, which were prepared by vaporizing the solvent in the solution of $C6PcH_2$ and $C6TBTAPH_2$, were heated to a temperature in the isotropic phase for injecting into the cell by capillary action, and cooled down to room temperature

IOP Conf. Series: Journal of Physics: Conf. Series **924** (2017) 012003 doi:10.1088/1742-6596/924/1/012003

in vacuum. A DC bias was applied between the electrodes using batteries connected in series, and the films were irradiated using a Nd:YAG laser (wavelength: 355 nm, pulse width: 1 ns). The transient current response was observed using an oscilloscope (Teledyne LeCroy HDO4054) in vacuum at room temperature, and the charge carrier mobility μ was calculated according to the equation $\mu = d^2/V\tau$, where d, V, and τ are the film thickness, applied bias, and transit time, respectively.

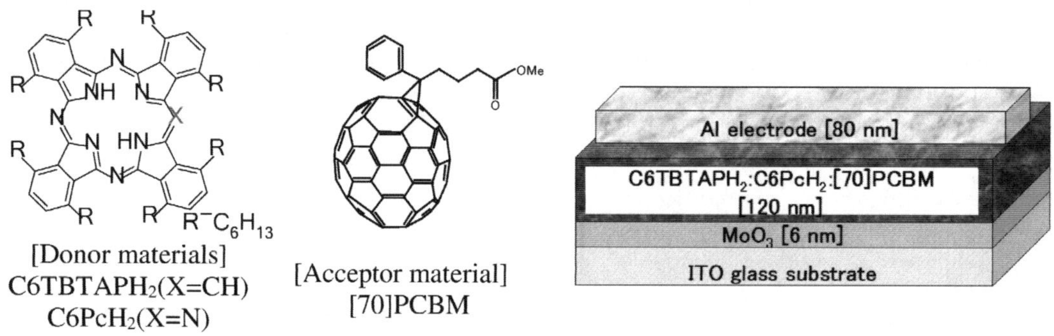

Figure 1. Molecular structures of $C6TBTAPH_2$, $C6PcH_2$, and [70]PCBM, and schematic device structure of the $C6TBTAPH_2$:$C6PcH_2$:[70]PCBM BHJ OSC.

3. Results and Discussion

Figure 2 shows the absorption spectra of $C6TBTAPH_2$:$C6PcH_2$:[70]PCBM thin films with various $C6PcH_2$ blend ratios. Two high absorption peaks at 300-450 nm and 600-800 nm were observed, which correspond to the optical transition at the B-band and Q-band of $C6TBTAPH_2$ or $C6PcH_2$, respectively. The absorbance of $C6TBTAPH_2$ at the B-band is higher than that of $C6PcH_2$, while the absorbance of $C6TBTAPH_2$ at the Q-band is lower than that of $C6PcH_2$ [15]. In the $C6TBTAPH_2$:$C6PcH_2$:[70]PCBM thin films, therefore, the absorbance at the B-band decreased, and that at the Q-band increased with increasing the blend ratio of $C6PcH_2$. It is noted that the marked change of the optical properties, such as absorption spectrum, appeared by adjusting the blend ratio of $C6TBTAPH_2$:$C6PcH_2$, in spite of the slight difference of the molecular structures.

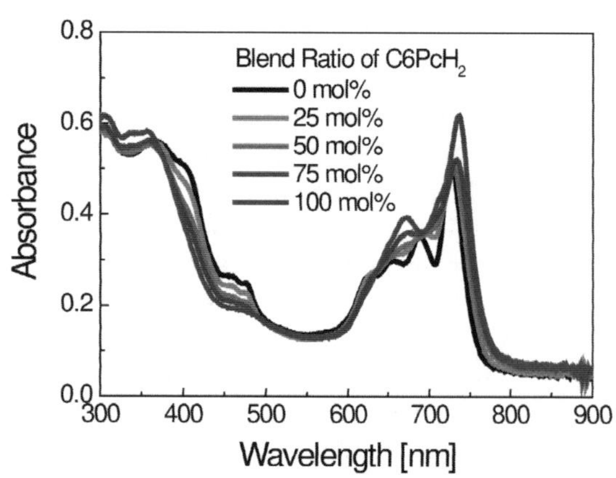

Figure 2. Absorption spectra of $C6TBTAPH_2$:$C6PcH_2$:[70]PCBM thin films with various $C6PcH_2$ blend ratios.

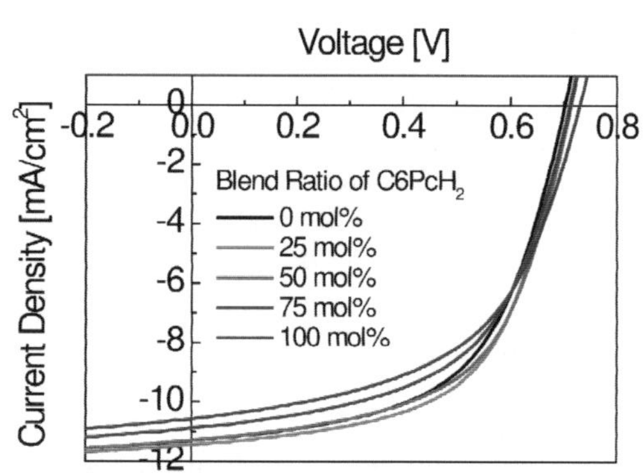

Figure 3. *J-V* characteristics of the $C6TBTAPH_2$:$C6PcH_2$:[70]PCBM BHJ OSC with various $C6PcH_2$ blend ratios.

The *J-V* characteristics of C6TBTAPH$_2$:C6PcH$_2$:[70]PCBM BHJ OSC are shown in Fig. 3, and the photovoltaic properties depending on the C6PcH$_2$ blend ratio are summarized in Table I. The open-circuit voltage (V_{oc}) of the solar cell with C6TBTAPH$_2$:[70]PCBM BHJ or C6PcH$_2$:[70]PCBM BHJ was 0.70 or 0.73 V, respectively. The V_{oc} of the solar cells with the blended donors of C6TBTAPH$_2$:C6PcH$_2$ were 0.71 V, which was close to that with C6TBTAPH$_2$:[70]PCBM BHJ. The V_{oc} of OSC is explained by an energy gap between the HOMO level of donors and the lowest occupied molecular orbital (LUMO) level of acceptors (ΔE_{DA}) and/or work function of metal electrodes [21, 22]. Although the difference between V_{oc} and ΔE_{DA} is still subject to debate, a linear relationship between V_{oc} and ΔE_{DA} has been experimentally confirmed [22]. The HOMO levels of C6TBTAPH$_2$ and C6PcH$_2$ were estimated to be 5.25 eV and 5.32 eV, respectively. Hence, the energy gap between the HOMO level of C6TBTAPH$_2$ and the LUMO level of [70]PCBM tends to be predominant for the V_{oc} of BHJ OSC with the blended donors of C6TBTAPH$_2$ and C6PcH$_2$. The fill factors (FF) of the solar cells with various blend ratios of C6PcH$_2$ were evaluated to be 0.56-0.58.

Table I. *J–V* characteristics of the C6TBTAPH$_2$:C6PcH$_2$:[70]PCBM BHJ OSC with various C6PcH$_2$ blend ratios.

Blend Ratio of C6PcH$_2$ (mol%)	V_{oc} (V)	J_{sc} (mA/cm^2)	FF	PCE (%)
0	0.70	11.3	0.57	4.5
25	0.71	11.6	0.58	4.8
50	0.71	11.3	0.58	4.6
75	0.71	10.9	0.56	4.3
100	0.73	10.6	0.53	4.1

The J_{sc} and PCE of C6TBTAPH$_2$:C6PcH$_2$:[70]PCBM BHJ OSC were markedly depending on the C6PcH$_2$ blend ratio as shown in Fig. 4. The J_{sc} and PCE of C6TBTAPH$_2$:[70]PCBM BHJ OSC were 11.3 mA/cm^2 and 4.5%, respectively, while those of C6PcH$_2$:[70]PCBM BHJ OSC were 10.6 mA/cm^2 and 4.1%, respectively. In the case of the C6TBTAPH$_2$:C6PcH$_2$:[70]PCBM BHJ OSC at 25 mol% of C6PcH$_2$ blend ratio, the J_{sc} and PCE were improved to be 11.6 mA/cm^2 and 4.8%, respectively. It is suggested that the enhancement of the J_{sc} mainly contribute to the improvement of the PCE.

Figure 5 shows the EQE spectra of C6TBTAPH$_2$:C6PcH$_2$:[70]PCBM BHJ OSC with various C6PcH$_2$ blend ratios. The main origins of the EQE peaks should be optical excitations at the B-band and Q-band of C6TBTAPH$_2$ or C6PcH$_2$. The EQE at B-band in the C6TBTAPH$_2$:[70]PCBM BHJ OSC is higher than that in the C6PcH$_2$:[70]PCBM BHJ OSC, while the EQE at Q-band in the C6TBTAPH$_2$:[70]PCBM BHJ OSC was lower than that in the C6PcH$_2$:[70]PCBM BHJ OSC. With increasing the blend ratio of C6PcH$_2$ in the C6TBTAPH$_2$:C6PcH$_2$:[70]PCBM BHJ OSC, the EQE at B-band decreased and that at Q-band slightly changed, which must be strongly related with the blend ratio dependence of the absorption spectra as abovementioned.

The EQE is defined as the ratio of the number of charge carriers collected at the electrodes in the solar cell to that of incident photons. In solar radiation spectrum of AM1.5, the number of photons around the wavelength of Q band is about two times more than that at B band. Comparing the EQE spectra at 0 mol% and 25 mol% of C6PcH$_2$ blend ratios, the quenching of the EQE at B-band with increasing the C6PcH$_2$ blend ratio seems to be more marked than the enhancement of the EQE at Q-band, however, the enhanced J_{sc} could be confirmed by integrating the EQE spectra indeed. The enhanced J_{sc} could be explained by the photon numbers used for the photon-electron conversion rather than the EQE. Therefore, the photosensitivity at the Q-band rather than the B-band should be still predominant for the device performance at 0 mol% and 25 mol% of C6PcH$_2$ blend ratios.

In the case of the C6PcH$_2$ blend ratio beyond 25 mol%, the absorbance at 400 nm monotonically decreased as shown in Fig. 6. With increasing the C6PcH$_2$ blend ratio, the absorption loss at B-band should be a main factor, which caused the decreasing of the PCE as well as J_{sc}. As a result, the highest PCE appeared at 25 mol% of C6PcH$_2$ blend ratio.

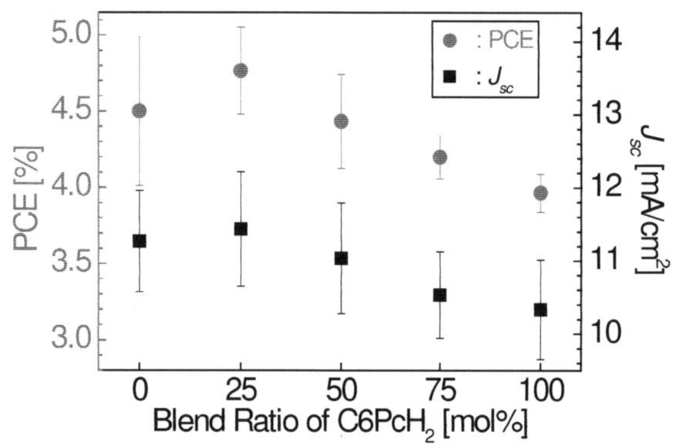

Figure 4. C6PcH$_2$ blend ratio dependence of the J_{sc} and PCE in the C6TBTAPH$_2$:C6PcH$_2$:[70]PCBM BHJ OSC.

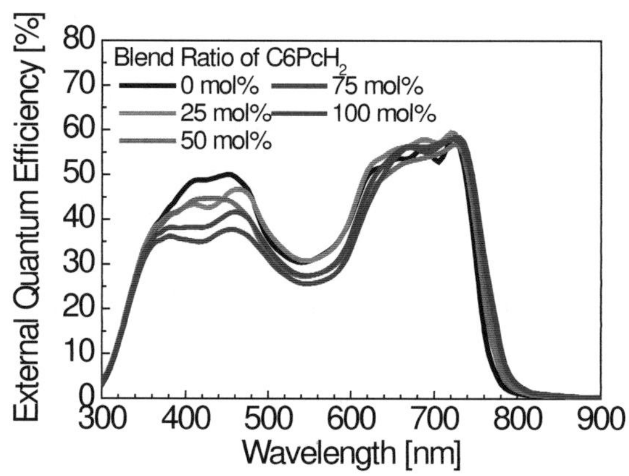

Figure 5. EQE spectra of C6TBTAPH$_2$:C6PcH$_2$:[70]PCBM BHJ OSC with various C6PcH$_2$ blend ratios.

In order to discuss about the carrier transport in the C6TBTAPH$_2$:C6PcH$_2$ blended film, the hole mobility as a fundamental property of the blend film was investigated by TOF technique. Figure 6 also shows the C6PcH$_2$ blend ratio dependence of hole mobility in the blended films of C6TBTAPH$_2$ and C6PcH$_2$. The hole mobility in the C6TBTAPH$_2$ film was 0.23 cm^2/Vs. In the blended films below 25 mol% of C6PcH$_2$ blend ratio, the mobility could not be evaluated, because it is hard to determine the transit time due to the dispersive decay curve with weak photocurrent. The hole mobility could be evaluated to be 0.47 cm^2/Vs at 25 mol% of C6PcH$_2$ blend ratio, and gradually enhanced with increasing the blend ratio of C6PcH$_2$, eventually 0.96 cm^2/Vs at 100 mol% of C6PcH$_2$. It is noted that the hole mobility changes depending on the blend ratio in spite of the slight difference on the molecular structures and that the blank region of the hole mobility exist below 25 mol% of C6PcH$_2$ blend ratio.

The crystal structures in the single crystal or thin film were reported to be based on a triclinic lattice for C6TBTAPH$_2$ and a monoclinic one for C6PcH$_2$ [23-25]. Further studies exhibited that the crystal structures in blended single crystals of C6TBTAPH$_2$ and C6PcH$_2$ at the C6PcH$_2$ blend ratio below 25 mol% were similar to the triclinic lattice of C6TBTAPH$_2$ [23]. In the range of 25-100 mol% of C6PcH$_2$ blend ratio, the crystal structures in the blended single crystals of C6TBTAPH$_2$ and C6PcH$_2$ were similar to the monoclinic lattice of C6PcH$_2$ [23]. It is considered, therefore, that the relatively high hole mobility originated from the C6PcH$_2$–based molecular packing structure [19].

As shown in Fig. 6, with increasing a blend ratio of C6PcH$_2$, the hole mobility was enhanced, while the absorbance at B-band was suppressed. At 25 mol% of C6PcH$_2$ blend ratio, the carrier generation induced by the photo-excitation at Q-band rather than B-band was predominant, and the hole mobility was higher than that of C6TBTAPH$_2$ because of the C6PcH$_2$–based crystal structure. Thus, the highest J_{sc} and PCE were achieved at 25 mol% of C6PcH$_2$ blend ratio.

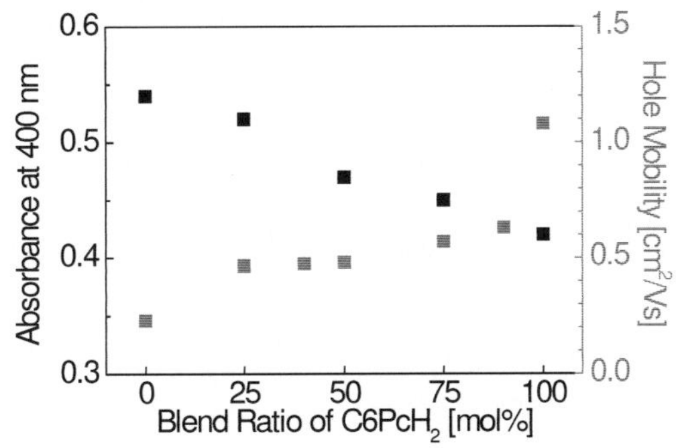

Figure 6. C6PcH_2 blend ratio dependence of the hole mobility in the C6TBTAPH_2:C6PcH_2 blended films and the absorbance at 400 nm of the C6TBTAPH_2:C6PcH_2:[70]PCBM BHJ films.

4. Conclusions

The charge carrier mobility in the blend films of C5PcH_2 and C6PcH_2 were investigated. The phase diagram of the mixture of C5PcH_2 and C6PcH_2 was determined from the thermal properties, it was found that the mixture possess the miscibility. In the case of C5PcH_2 blend ratio below 25 mol%, both hole and electron mobilities were kept high values comparable with those of C6PcH_2, while both hole and electron mobilities markedly decreased by one order of magnitude in the case of C5PcH_2 blend ratio above 30 mol%. From the results of XRD measurement, the molecular structure of the mixture film obviously changed depending on the blend ratio of C5PcH_2, that must be strongly related with the behavior of the charge carrier transport.

Acknowledgements

This work was partly supported by Advanced Low Carbon Technology Research and Development Program from the Japan Science and Technology Agency (JST-ALCA), by JSPS KAKENHI Grant Numbers 15H03552, 26600073, 24246009, and by JSPS Core-to-Core Program, A. Advanced Research Networks.

References

[1] Schmidt-Mende L, Fechtenkotter A, Mullen K, Moons E, Friend R H and MacKenzie J D 2001 *Science* **293** 1119
[2] Poll T S, Love J A, Nguyen T Q and Bazan G C 2012 *Adv. Mater.* **24** 3646
[3] Kyaw A K K, Wang D H, Wynands D, Zhang J, Nguyen T Q, Bazan G C and Heeger A J 2013 *Nano Lett.* **13** 3796
[4] Walker B, Kim C and Nguyen T Q 2011 *Chem. Mater.* **23** 470
[5] O'Neil M and Kelly S M 2011 *Adv. Mater.* **23** 566
[6] Shimizu Y, Oikawa K, Nakayama K and Guilon D 2007 *J. Mater. Chem.* **17** 4223
[7] Miyake Y, Shiraiwa Y, Okada K, Monobe H, Hori T, Yamasaki N, Yoshida H, Cook M J, Fujii A, Ozaki M and Shimizu Y 2011 *Appl. Phys. Express* **4** 0214604
[8] Kaafarani B R 2011 *Chem. Mater.* **23** 378
[9] Sergeyev S, Pisula W and Geerts Y H 2007 *Chem. Soc. Rev.* **36** 1902
[10] Dao Q D, Hori T, Fukumura K, Masuda T, Kamikado T, Fujii A, Shimizu Y and Ozaki M 2012 *Appl. Phys. Lett.* **101** 263301
[11] Matsuo Y, Sato Y, Niinomi T, Soga I, Tanaka H and Nakamura E 2009 *J. Am. Chem. Soc.* **131** 16048

IOP Conf. Series: Journal of Physics: Conf. Series **924** (2017) 012003 doi:10.1088/1742-6596/924/1/012003

[12] Guide M, Lin J D A, Proctor C M, Chen J, García-Cervera C and Nguyen T Q 2014 *J. Mater. Chem. A* **2** 7890
[13] Dao Q D, Hori T, Fukumura K, Masuda T, Kamikado T, Fujii A, Shimizu Y and Ozaki M 2013 *Org. Electron.* **14** 2628
[14] Dao Q D, Watanabe K, Itani H, Sosa-Vargas L, Fujii A, Shimizu Y and Ozaki M 2014 *Chem. Lett.* **43** 1761
[15] Dao Q D, Sosa-Vargas L, Higashi T, Ohmori M, Itani H, Fujii A, Shimizu Y and Ozaki M 2015 *Org. Electron.* **23** 44
[16] Fukui H, Nakano S, Uno T, Dao Q D, Saito T, Fujii A, Shimizu Y and Ozaki M 2014 *Org. Electron.* **15** 1189
[17] Wegewijs B R and Siebbeles L D A 2002 *Phys. Rev. B* **65** 245112
[18] Schweicher G, Gbabode G, Quist F, Debever O, Dumont N, Sergeyev S and Geerts Y H 2009 *Chem. Mater.* **21** 5867
[19] Nakagawa D, Nakano C, Ohmori M, Fujii A and Ozaki M 2017 *Org. Electron.* **44** 67
[20] Fujii A, Itani H, Watanabe K, Dao Q D, Sosa-Vargas L, Shimizu Y and Ozaki M to be published in *Mol. Cryst. & Liq. Cryst.*
[21] Mihailetchi V D, Blom P W M, Hummelen J C and Rispens M T 2003 *J. Appl. Phys.* **94** 6849
[22] Yamamoto S, Orimo A, Ohkita H, Benten H and Ito S 2012 *Adv. Energy Mater.* **2** 229
[23] Nakano C, Ohmori M, Fujii A, Tohnai N and Ozaki M *J. Cryst. Growth* in press
[24] Ohmori M, Nakano C, Miyano T, Tohnai N, Fujii A and Ozaki M 2016 *J. Cryst. Growth* **445** 9
[25] Ohmori M, Nakano C, Fujii A, Shimizu Y and Ozaki M *J. Cryst. Growth* in press

IOP Conf. Series: Journal of Physics: Conf. Series **924** (2017) 012004 doi:10.1088/1742-6596/924/1/012004

Aggregation States of Organic Dye Molecules in Organic – Inorganic Hybrid Films Studied by Near-Field and Far-Field Fluorescence Spectroscopy

M. Ishihara

Hyogo Prefectural Institute of Technology, 3-1-12 Yukihira-cho, Suma-ku, Kobe, 654-0037, Japan

E-mail: mari@hyogo-kg.jp

Abstract. Near-field fluorescence spectroscopy in conjunction with far-field fluorescence and Raman spectroscopy has been applied to investigate the aggregation states of cyanine dye of 2-[5-(1,3-Dihydro-3,3-dimethyl-1-octadecyl-2H-indol-2-ylidene)-1,3-pentadienyl]-3,3-dimethyl-1-octadecyl-3H-indolium perchlorate (NK3175) molecules in the depth direction, from the nanometer-scale top surface to the bulk, in the hybrid films consisted of NK3175 and a clay compound (SWN). The far-field Raman spectra reveal that these hybrid films contained both NK3175 molecules which did not take part in the adsorption and those adsorbed on SWN. The near-field photoluminescence (PL) spectra in the outermost surface within ca. 30 nm exhibit a significant blue shift for the hybrid films as compared to their respective PL spectra in the bulk, implying that the top surface of these hybrid films becomes more polar and/or rigid in terms of the microenvironment around NK3175 molecules compared to the bulk of them, in contrast to the case for the other hybrid film which was prepared by the alternative method. These results provide new information on the aggregation states of organic dye molecules not only in the bulk but in the nanometer-scale top surface of organic-inorganic hybrid systems.

1. Introduction

Photoinduced spectral change such as optical reflectance, refractivity, and fluorescence of photo-functional organic dyes has been utilized for various optical devices, for example, memories and switching devices. In order to find out the method for the non-destructive readout which is one of the most pressing problems for photochromic memories [1], we proposed to monitor the absorption in the UV-visible region for the cooperative responsive system of a photochromic dye and a cyanine dye [2]. In this system, the photoinduced spectral change ascribed to a cyanine dye (NK3175) was generated by the hybridization of NK3175, a photochromic dye and an inorganic layered material of cation-exchangeable clay, smectite (SWN), with a simple preparation method [2]. Furthermore, we have found that the spectral change due to NK3175 was induced upon the irradiation of UV light without the help of photochromism, for the first time, by the hybridization of NK3175 and SWN [3]. Inorganic layered materials such as clay minerals are extensively utilized as host materials for functional organic-inorganic hybrid systems due to their characteristic properties of intercalation and adsorption of organic guest species such as organic dyes [4-9]. It has been reported that intercalation and adsorption of organic dyes promote aggregation and alignment of them in hybrid systems consisted of organic dyes and inorganic layered materials [4-5, 10-16]. The aggregation states of cyanine dyes which are the best known self-aggregation dyes have been studied widely for the organic-inorganic hybrid systems [17]. In the above film hybridized with NK3175 and SWN the aggregation states of NK3175 molecules and the interaction between NK3175 and SWN might affect the microstructure and photoinduced spectral behavior. In this point of view it is of interest to investigate the aggregation states of NK3175 molecules in the above film.

It is well-known that the spectral properties of cyanine dyes are influenced by their aggregation [18]. Thus far, the structure of molecular assemblies and the aggregation states of cyanine dyes organized on silver halides [19-20] and clays [10, 21], etc. has been studied extensively by using spectroscopic methods such as electronic absorption and fluorescence spectroscopy. Near-field scanning optical spectroscopy (NSOM), which has been developed recently, is high resolution probe techniques on nanometer length scale beyond the diffraction limit [22]. Near-field fluorescence spectroscopy, one of these techniques, has been used to study nanostructured thin-film materials, including aggregated thin films of fluorescent molecules [23-25] and locally aggregated conjugated polymer thin films [26-27], which has probed the local molecular orientation and interaction. Near-field fluorescence spectra collected photoluminescence from the nanometer-scale surface of films which are excited in the near-field ("illumination-collection mode") certainly reflect chemical properties only in the nanometer-scale top surface of films. Consequently, simultaneous use of near- and far-field fluorescence spectroscopy is expected to give new information on the aggregation states of cyanine dye molecules not only in the bulk but in the nanometer-scale top surface in their organization.

In this study, near-field fluorescence spectroscopy in conjunction with far-field fluorescence and Raman spectroscopy has been applied to investigate the aggregation states of NK3175 molecules in the depth direction in several kind of hybrid films consisted of NK3175 and SWN, which might provide new information on the mechanism of the photoinduced spectral change due to NK3175 for the above film. Near-field photoluminescence (PL) spectra along with far-field ones reveal that the aggregation states of NK3175 molecules in the top surface of the above film are different from those in the bulk, in contrast to the case in the other hybrid film prepared by the alternative method. This study is expected to demonstrate new application of simultaneous use of near- and far- field PL spectroscopy to the characterization of films in the depth direction, from a nanometer-scale top surface to the bulk.

2. Experimental

2.1. Materials

A cyanine dye of 2-[5-(1,3-dihydro-3,3-dimethyl-1-octadecyl-2H-indol-2-ylidene)-1,3-pentadienyl]-3,3-dimethyl-1-octadecyl-3H-indolium perchlorate (NK3175; Hayashibara Biochemical Labs., Inc.) was used without further purification (Figure 1). Cation-exchangeable clay, SWN (CO-OP Chemical Co., Inc.), was used as a host material. It is a type of hydrophilic smectite, and has good transparency in the visible region. The composition is $[(Mg_{2.67}Li_{0.33})(Si_4O_{10})(OH)_2]Na_{0.33}$, and the cation exchangeable capacity (CEC) is c.a. 0.6 $meq{\cdot}g^{-1}$. Distilled and deionized water (conductivity $< 4 \times 10^{-6}$ $S{\cdot}cm^{-1}$) and ethanol (Wako Pure Chemical Industries Co. Inc., spectroscopic grade) were used as solvents for the preparation of the organic-inorganic hybrid films.

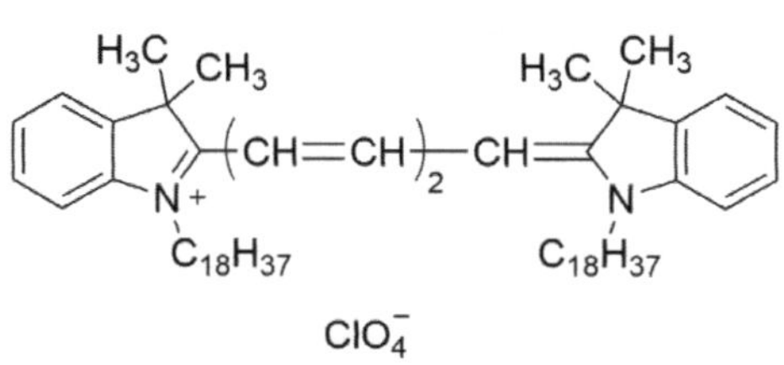

Figure 1. Chemical structure of cyanine dye of NK3175.

2.2. Preparation of the hybrid films

Solution of cyanine dye of NK3175 (1.5×10^{-6} mol) in ethanol (10 mL) added to aqueous suspensions of SWN (0.050 g/5mL) stirred at 60°C for 27 h in the dark. Aqueous suspensions of SWN were made by sonicating the mixture of SWN and deionized water for 60 min. or 5 min. Obtained viscous solutions were then casted on fused silica plates and dried at room temperature and transparent blue films were prepared. The obtained organic-inorganic hybrid films are, hereafter, referred to as NK3175-SWN(60min) and NK3175-SWN(5min). Another film of NK3175 and SWN, NK3175-SWN(dipping), was made by the alternative preparation method as follows. An aqueous suspension of SWN (0.020 g/5mL) was casted on a fused silica plate and dried at room temperature for 48 h and at 100°C for 3 h. Obtained SWN film was dipped in an ethanol solution of NK3175(1.5×10^{-4} M) at room temperature for 3.8 h, rinsed with ethanol, and dried at room temperature and a transparent blue film was prepared.

2.3. Measurement

Near- and far-field photoluminescence (PL) spectra were measured on near-field scanning optical spectrometer combined with microscopic optical spectrometer of NFS-230HGK (JASCO). Far-field Raman spectra were taken by laser Raman spectrometer of NRS-5100 (JASCO). An excitation wavelength of 532 nm produced by a Nd:YVO_4 laser was used for all far- and near-field measurement. The laser beam was produced a ca. 1 μm diameter spot on the sample for far-field measurement, thus, spatial resolution could be roughly estimated to be ca. 1μm. Near-field PL measurement was performed in the illumination-collection mode and using double-tapered fiber probes with ca. 300, 180, 100, and 30 nm diameter apertures (JASCO Co., Inc.). According to previous reports [28-30] that the maximum spatial resolution of near-field scanning optical spectrometry (NSOM) is given by the diameter of the aperture, spatial resolution not only in horizontal direction but also in depth direction was supposed to be the aperture size of a fiber probe. Each near-field PL measurement was carried out without changing the measuring position of the sample which was set on far-field measurement. The near- and far-field PL profiles were reproducible and hardly changed from position to position and photodamage to the samples was not observed.

3. Results and discussion

3.1. Far-field Raman spectra of the hybrid films

Spectral change due to NK3175 in the visible region was occurred upon UV light irradiation in NK3175-SWN(60min) in agreement with our previous work [3], while that was hardly recognized in NK3175-SWN(5min) and NK3175-SWN(dipping). There might be a possibility that the comparison of the interaction between NK3175 and SWN and the aggregation states of NK3175 molecules among NK3175-SWN(60min), NK3175-SWN(5min) and NK3175-SWN(dipping) could give useful information on the mechanism of photoinduced spectral change in NK3175-SWN(60min). In the first place the far-field Raman measurement was performed for NK3175-SWN(60min), NK3175-SWN(5min) and NK3175-SWN(dipping). Figure 2 shows the far-field Raman spectra for the above hybrid films consisted of NK3175 and SWN and, for reference, the NK3175 powder. The observed bands at 1603, 1501, 1470, 1406, 1363 and 928 cm^{-1} for the above hybrid films (Figure 2a, 2b, and 2c) were in accord with those for NK3175 powder (Figure 2d), indicating that there are NK3175 molecules which did not participate in the chemical interaction with SWN such as adsorption in the above hybrid films. According to previous reports [31- 32], the band at 1603 cm^{-1} is assignable to C=N stretching motion of the end chromophore of NK3175 and the bands at 1470, 1406 and 1363 cm^{-1} might contain the contribution from the end chromophore vibration of NK3175. On the other hand, the new bands at 1556, 1433, 1300, 1173, 1124, and 1021 cm^{-1} appeared in the far-field Raman spectra for the above hybrid films, which is different from the far-field Raman spectrum for NK3175 powder (Figure 2d), and the bands at 1470 cm^{-1} for NK3175-SWN(60min) (Figure 2a) and NK3175-

SWN(5min)(Figure 2b) were relatively smaller than that for NK3175 powder. The XRD patterns of the above hybrid films in which the 100 reflection were observed around d = 14.7 Å were almost similar to that of the SWN film in accord with our previous work [3], indicating that NK3175 molecules were not intercalated into SWN layers in the above hybrid films. These results are attributable to the adsorption of the cationic dye of NK3175 on the external anionic surface of SWN in agreement with our previous work [3], although the features of the above near-field Raman spectra do not seem to correspond to the photoinduced spectral behaviour for the above hybrid films. In Figure 2c the new bands at 1556, 1433, 1300, 1173, 1124, and 1021 cm^{-1} were relatively smaller than those in Figure 2a and 2b, suggesting that for NK3175-SWN(dipping) the number of NK3175 molecules adsorbed on SWN relative to NK3175 molecules which are independent of adsorption is smaller in comparison with that for NK3175-SWN (60min) and NK3175-SWN(5min). This might be originated from the difference in the preparation methods of these hybrid films. A possible explanation is as follows; for NK3175-SWN(60min) and NK3175-SWN(5min) NK3175 molecules and SWN were sufficiently mixed in liquid phase to result in the adequate adsorption of NK3175 molecules on SWN. On the contrary, in NK3175-SWN(dipping) there might be no appreciable chemical interaction between NK3175 molecules and SWN hardly to occur the adsorption of NK3175 molecules on SWN because of the introduction of NK3175 molecules not into the liquid phase but into the solid phase of SWN.

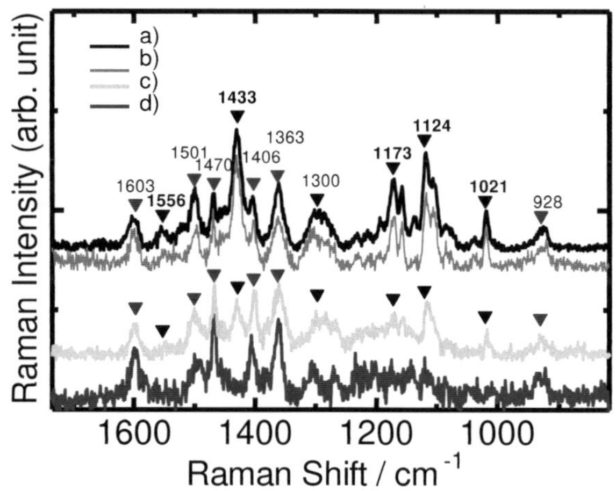

Figure 2. Far-field Raman spectra of a) NK3175-SWN(60min), b) NK3175-SWN(5min), c) NK3175-SWN(dipping), and d) NK3175 powder. ▼ indicates the bands observed for NK3175 powder, whereas ▼ indicates the newly observed bands for the above hybrid films.

3.2. Far-field PL spectra of the hybrid films

The position of the PL maxima is known to reflect the nature of the surrounding environment which affects aggregation states and is explained in terms of polar and/or rigid nature of the microenvironment. Figure 3 shows the far-field PL spectra of the hybrid films consisted of NK3175 and SWN and, for reference, the NK3175 film spin-coated from a THF solution of NK3175, referred to as NK3175 film. NK3175-SWN(60min) and NK3175-SWN(5min) exhibited their PL maxima around 700 nm with the shoulder-like structure around 650 nm, whereas for NK3175-SWN(dipping) the PL maximum was slightly red-shifted to ca. 709 nm and this band shape was rather unsymmetrical compared to that for NK3175-SWN(60min) and NK3175-SWN(5min). This unsymmetrical PL shape might be due to the self-absorption of the fluorescence of NK3175 molecules in NK3175-SWN(dipping). Consequently, the far-field PL spectra of NK3175-SWN(60min), NK3175-SWN(5min), and NK3175-SWN(dipping) are fundamentally similar to that of NK3175 film, which indicates that the aggregation states of NK3175 molecules roughly resemble in these films. The electronic absorption spectra of NK3175-SWN(60min) [3], NK3175-SWN(5min), and NK3175-SWN(dipping) were similar to show the absorption bands around 650 and 603 nm probably ascribed to monomers and H-dimers of NK3175, respectively [10, 17], which is in accordance with the above far-field PL result that the aggregation states of NK3175 molecules approximately resemble in these films. It has been reported that the blue shift of the PL maximum indicates polarity and/or rigidity of the microenvironment of organometallic complex ions of $Ru(bpy)_3^{2+}$ in the hybrid systems of these ions and inorganic layered materials [33]. From this fact and nonfluorescent ability of H-dimers, for NK3175-SWN(60min) and NK3175-SWN(5min) it could be assumed that the shoulder-like structure around 650 nm which was slightly appeared in the far-field PL spectra is blue-shifted from the PL maxima around 700 nm based on monomers of NK3175 and that this shoulder-like structure is possibly ascribed to the polar and/or rigid environment around NK3175 molecules. In analogy with the far-field Raman results, the features of the far-field PL and the electronic absorption spectra for the above hybrid films do not seem to correspond to their photoinduced spectral behaviour.

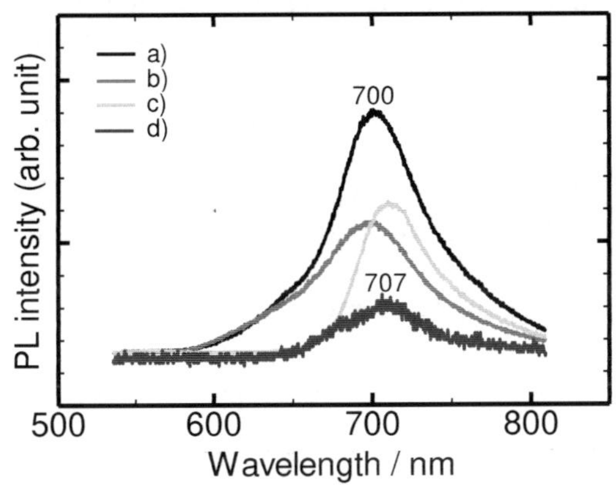

Figure 3. Far-field photoluminescence spectra of a) NK3175-SWN(60min), b) NK3175-SWN(5min), c) NK3175-SWN(dipping) , and d) NK3175 film.

3.3. Near-field PL spectra of the hybrid films

The near-field PL spectra for NK3175-SWN(60min) and NK3175-SWN(5min) were dependent on the aperture diameters of fiber probes used on their measurement in contrast to those for NK3175-SWN(dipping). The features of near-field PL spectra for the above hybrid films do not seem to correspond to their photoinduced spectral behaviour as well. Further investigation should be needed in order to clarify the mechanism of the phtoinduced spectral change in NK3175-SWN(60min). Figures 4, 5 and 6 show the near-field PL spectra for NK3175-SWN(60min), NK3175-SWN(5min) and NK3175-SWN(dipping), respectively. The observed near- and far-field PL maxima are summarized in Table 1. For NK3175-SWN(60min) the broad near-field PL maxima were observed around 700 nm in the use of fiber probes with ca. 300, 180, and 100 nm diameter apertures (Figure 4a, 4b, and 4c) and the fluorescent shoulder which might be correspond to the shoulder-like structure slightly appeared around 650 nm in its far-field PL spectrum (Figure 3a) was also observed around 650 nm in the use of a fiber probe with a ca. 300 nm diameter aperture (Figure 4a). These near-field PL profiles were nearly similar to the far-field PL spectrum of NK3175-SWN(60min). These results suggest that the aggregation states of NK3175 molecules in the inside more than ca. 100 nm from the top surface of NK3175-SWN(60min) are almost analogous to those in the bulk. On the other hand, the near-field PL maximum in the use of a fiber probe with a ca. 30 nm diameter aperture (Figure 4d) was significantly blue-shifted to ca. 649 nm and the half-value width of this band was relatively narrow compared to the near-field PL spectra taken by fiber probes with larger diameter apertures (Figure 4a, 4b, and 4c). This indicates that the aggregation states of NK3175 molecules in the outermost surface of NK3175-SWN(60min) within ca. 30 nm are different from those in the bulk and are somewhat limited. In this way it is demonstrated that the near-field PL spectra measured by using fiber probes with various aperture sizes together with the far-field ones give fruitful information on the aggregation states of

NK3175 molecules in NK3175-SWN(60min) not only in the bulk but in the top surface with the level of several tens of nanometers.

The near-field PL spectra for NK3175-SWN(5min) exhibited the similar tendency to those for NK3175-SWN(60min) that the aggregation states of NK3175 molecules in the outermost surface within ca. 30 nm of NK3175-SWN(5min) are different from those in the bulk. The broad near-field PL bands were observed around 694 nm by using fiber probes with ca. 300 and 180 nm diameter apertures (Figure 5a and 5b). On the other hand, the near-field PL maximum in the use of the fiber probe with a ca. 30 nm diameter aperture (Figure 5c) was considerably blue-shifted to ca. 649 nm and the half-value width of this band was relatively narrow compared to the near-field PL spectra taken by fiber probes with larger diameter apertures (Figure 5a and 5b). According to the previous report [33], it could be assumed that the polar and/or rigid environment around NK3175 molecules is greatly induced in the top surface of NK3175-SWN(60min) and NK3175-SWN(5min), leading to the blue shift in their near-field PL spectra. On the other hand, this kind of environment around NK3175 molecules is presumed to be slightly produced in the bulk of NK3175-SWN(60min) and NK3175-SWN(5min). It is well-known that the structure of a solid surface is generally different from that in the bulk and influences many of its important properties such as chemical and electronic properties. Accordingly, for NK3175-SWN(60min) and NK3175-SWN(5min) the arrangement of NK3175 molecules and SWN in the top surface could be somewhat different from that in the bulk. One of the possible situations is that there is more number of NK3175 molecules externally adsorbed on SWN in the top surface relative to those in the bulk of NK3175-SWN(60min) and NK3175-SWN(5min). In this case, the more polar and/or rigid microenvironment of NK3175 molecules could be caused in the top surface of these hybrid films, resulting in the blue-shifted photoluminescence.

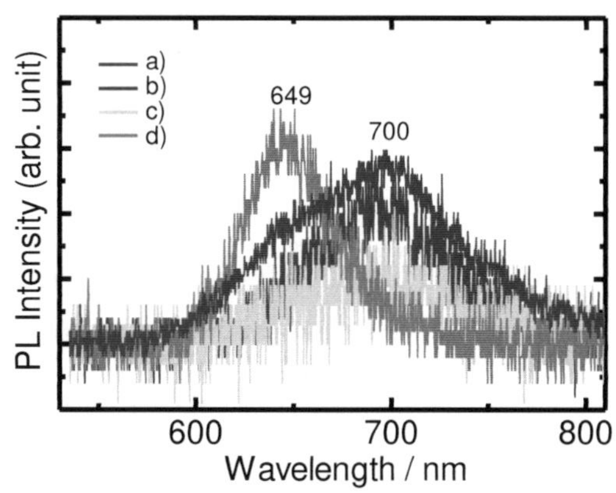

Figure 4. Near-field photoluminescence spectra of NK3175-SWN(60min) measured by using fiber probes with various opening diameters of a) 300 nm, b) 180 nm, c) 100 nm, and d) 30 nm.

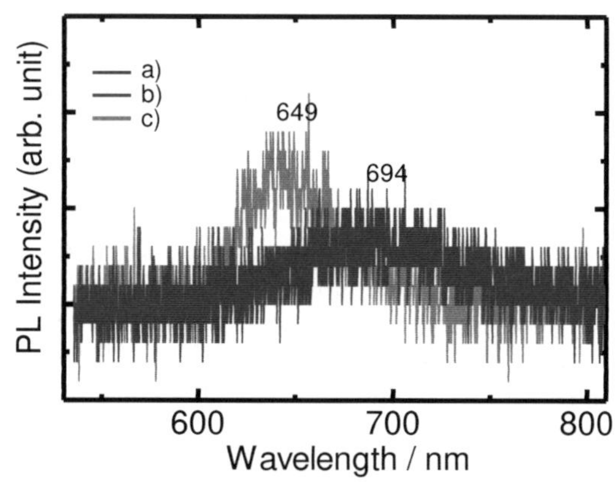

Figure 5. Near-field photoluminescence spectra of NK3175-SWN(5min) measured by using fiber probes with various opening diameters of a) 300 nm, b) 180 nm, and c) 30 nm.

On the other hand, the near-field PL spectra for NK3175-SWN(dipping) taken by using fiber probes with ca. 100 and 30 nm diameter apertures were similar. Moreover, these PL maxima at ca. 704 nm were comparable to the far-field PL spectrum of NK3175-SWN(dipping). These results indicate that the aggregation states of NK3175 molecules in the outermost surface within ca. 30 nm of NK3175-SWN(dipping) are in accordance with those in the bulk, in contrast to the cases in NK3175-SWN(60min) and NK3175-SWN(5min). These difference in the aggregation states of NK3175 molecules in the top surface between NK3175-SWN(dipping) and the other hybrid films of NK3175-

SWN(60min) and NK3175-SWN(5min) might come from the preparation method of these films, although the details of this origin are not clear.

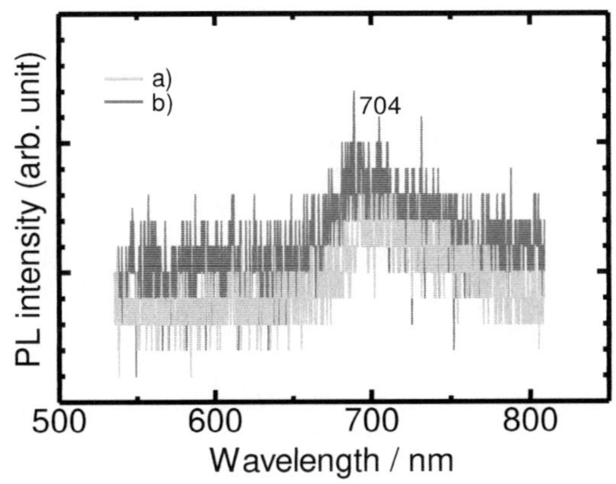

Figure 6. Near-field photoluminescence spectra of NK3175-SWN(dipping) measured by using fiber probes with opening diameters of a) 100 nm and b) 30 nm.

Table 1. Observed near- and far-field photoluminescence (PL) maxima for the hybrid films of NK3175 and SWN.

Samples	Observed PL maxima / nm				
	opening diameters / nm				diameter of a beam spot / μm
	30	100	180	300	1
NK3175-SWN (60min)	649	696	698	700	700
NK3175-SWN (5min)	649	----------	694	694	699
NK3175-SWN (dipping)	704	704	----------	----------	709

4. Conclusions

The aggregation states of NK3175 molecules in the hybrid films consisted of NK3175 and SWN, NK3175-SWN(60min), NK3175-SWN(5min) and NK3175-SWN(dipping), not only in the bulk but in the outermost surface within ca. 30 nm are demonstrated by simultaneous use of near- and far-field PL spectra. The aggregation states of NK3175 molecules in the outermost surface within ca. 30 nm of NK3175-SWN(60min) and NK3175-SWN(5min) are different from those in the bulk, in contrast to the case in NK3175-SWN(dipping) which was made by the alternative preparation method. The blue-shifted near-field photoluminescence for NK3175-SWN(60min) and NK3175-SWN(5min) suggests the top surface of these films to become more polar and/or rigid in terms of the microenvironment around NK3175 molecules relative to the bulk of them. These results are expected to show new application of simultaneous use of near- and far- field PL spectra to the characterization of films in the depth direction, from the top surface with the level of several tens of nanometers to the bulk. The far-field Raman spectra exhibit that the above hybrid films contained both NK3175 molecules which did not take part in the adsorption and those adsorbed on the external surface of SWN.

Acknowledgements

We would like to thank Dr. Nobutaka Tanigaki of National Institute of Advanced Industrial Science and Technology (AIST) for his valuable discussions and advices.

References

[1] Irie M 2000 *Chem. Rev.* **100** 1685
[2] Ishihara M, Hirase R, Mori M, Yoshioka H and Ueda Y 2009 *Thin Solid Films* **518** 857

[3] Ishihara M, Hirase R and Yoshioka H 2014 *World Journal of Nano Science and Engineering* **4** 126
[4] Ogawa M and Kuroda K 1995 *Chem. Rev.* **95** 399
[5] Shichi T and Takagi K 2000 *J. Photochem. Photobiol. C* **1** 113
[6] Takagi S, Shimada T, Ishida Y, Fujimura T, Masui D, Tachibana H, Eguchi M and Inoue H 2013 *Langmuir* **29** 2108
[7] Zhou C H, Shen Z F, Liu L H and Liu S M 2011 *J. Mater. Chem.* **21** 15132
[8] Ras R H A, Umemura Y, Johnston C T, Yamagishi A and Schoonheydt R A 2007 *Phys. Chem. Chem. Phys.* **9** 918
[9] Ogawa M, Ishii I, Miyamoto N and Kuroda K 2003 *Appl. Clay Sci.* **22** 179
[10] Bujdak J 2006 *Appl. Clay Sci.* **34** 58
[11] Estevez M J T, Arbeloa F L, Arbeloa T L and Arbeloa I L 1994 *J. Colloid Interface Sci.* **162** 412
[12] Yariv S, Nasser A and Baron P 1990 *J. Chem. Soc., Faraday Trans.* **86** 1593
[13] Lucia L A, Yui T, Sasai R, Yoshida H, Takagi S, Takagi K, Whitten D G and Inoue H 2003 *J. Phys. Chem. B* **107** 3789
[14] Takagi K, Kurematsu T and Sawaki Y 1991 *J. Chem. Soc., Perkin Trans. 2* 1517
[15] Sasai R, Ogiso H, Shindachi I, Shichi T and Takagi K 2000 *Tetrahedron* **56** 6979
[16] Iyi N, Kurashima K and Fujita T 2002 *Chem. Mater.* **14** 583
[17] Mishra A, Behera R K, Behera P K, Mishra B K and Behera G B 2000 *Chem. Rev.* **100** 1973
[18] Jelly E E 1936 *Nature* **138** 1009
[19] Muenter A A, Brumbaugh D V, Apolito J, Horn L A, Spano F C, Mukamel S 1992 *J. Phys. Chem.* **96** 2783
[20] Maskasky J E 1991 *Langmuir* **7** 407
[21] Ogawa M, Kawai R and Kuroda K 1996 *J. Phys. Chem.* **100** 16218
[22] Betzig E and Trautman J K 1992 *Science* **257** 189
[23] Higgins D A and Barbara P F 1995 *J. Phys. Chem.* **99** 3
[24] Vanden Bout D A, Kerimo J, Higgins D A and Barbara P F 1996 *J. Phys. Chem.* **100** 11843
[25] Kerimo J, Adams D M, Barbara P F, Kaschak D M and Mallouk T E 1998 *J. Phys. Chem. B* **102** 9451
[26] Reid P J, Higgins D A and Barbara P F 1996 *J. Phys. Chem.* **100** 3892
[27] Schaller R D, Lee L F, Johnson J C, Haber L H, SayKally R J, Vieceli J, Benjamin I, Nguyen T-Q and Schwartz B J 2002 *J. Phys. Chem. B* **106** 9496
[28] Moerner W E, Plakhotnik T, Irngartinger T, Wild U P, Pohl D W and Hecht B 1994 *Phys. Rev. Lett.* **73** 2764
[29] Lieberman K, Harush S, Lewis A and Kopelman R 1990 *Science* **247** 59
[30] Knoll B and Keilmann F 1999 *Nature* **399** 134
[31] Yang J-P and Callender R H 1985 *J. Raman Spectrosc.* **16** 319
[32] Sato H, Kawasaki M, Kasatani K and Katsumata M 1988 *J. Raman Spectrosc.* **19** 129
[33] Nakato T, Kusunoki K, Yoshizawa K, Kuroda K and Kaneko M 1995 *J. Phys. Chem.* **99** 17896

Fabrication of Carbon Nanotube Polymer Actuator Using Nanofiber Sheet

Hayato Kato, Akikazu Shimizu, Taiga Sato and Masahito Kushida

Graduate School of Engineering, Chiba University, 1-33, Yayoi-cho, Inage-ku, Chiba 263-8522, Japan

E-mail: kushida@faculty.chiba-u.jp

Abstract. Carbon nanotube polymer actuators were developed using composite nanofiber sheets fabricated by multi-walled carbon nanotubes(MWCNTs) and poly (vinylidene fluoride-co-hexafluoropropylene) (PVDF-HFP). Nanofiber sheets were fabricated by electrospinning method. The effect of flow rate and polymer concentration on nanofiber formation were verified for optimum condition for fabricating nanofiber sheets. We examined the properties of MWCNT/PVDF-HFP nanofiber sheets, as follows. Electrical conductivity and mechanical strength increased as the MWCNT weight ratio increased. We fabricated carbon nanotube polymer actuators using MWCNT/PVDF-HFP nanofiber sheets and succeeded in operating of our actuators.

1. Introduction

Recently, soft materials that have the ability to convert electrical energy directly into mechanical work [1-3] have attracted a lot of attention [4] because it is expected to be applicable to robots, tactile and optical display, prosthetic devices, medical devices [5] and microelectromechanical system [6]. Among these soft materials, low-voltage driven electroactive polymer (EAP) actuators that can operate quickly and softly are focused on [7] due to lightness [8], easy of fabrication, and flexibility. Therefore, the EAP actuators can be used as artificial muscle [9] for various bio-medical and human affinity applications. Although ordinary EAP actuators operate only in electrolyte solutions [10], Ding et al. reported 'dry' EAP actuators that work in air, using ionic liquids (ILs) and conductive polymer [11]. ILs have low volatility [12], exhibit high ionic conductivities [13] and wide potential windows, which are advantageous for quick response actuators and high electrochemical stability components. However, conductive polymer-based actuators have essential drawback in their lifetime and responsiveness, because the motion is driven via oxidation/reduction of the conjugated polymers [6]. There were also some reports about actuators using single-walled carbon nanotube (SWCNT) [5, 14] and these actuators are driven via electrical charge/discharge of SWCNTs [6]. However, the actuation occurs only in electrolyte solutions. Therefore, carbon nanotube polymer actuators using ILs were proposed [2, 4, 15, 16]. Carbon nanotube (CNT) has advantage of superb electrical conductivity [17], surface width [18], and mechanical strength [19]. Therefore, many potential applications have been proposed for CNT such as conductive and high strength composites, energy storage and conversion devices, and semiconductor devices [20]. Mukai *et al.* reported the mechanism of this actuator using ILs. as following: when voltage is applied to the electrodes, the internal ions move and then EAP actuator is caused a volume change between the positive and negative electrodes. As a result, the actuator bends [2]. Carbon nanotube polymer actuators are currently being studied aiming at further improvement of the performance such as displacement and response speed in order to apply actuator to various fields [21]. Achieving this improvement, nanofiber sheet fabricated with conjugated polymer was proposed. Increasing the interfacial area between the electrolyte and the conjugated polymer should lead to a larger volume change of electrode films. In order to increase the interfacial area, nanofiber sheet of conductive polymer which has large specific surface area was applied to actuator's electrodes [22].

Therefore, we propose to fabricate carbon nanotube polymer actuator using electrolyte film containing ILs and multi-walled carbon nanotube (MWCNT) composite nanofiber electrode films which has larger specific surface area than bulk film to improve furthermore actuator performance. We used MWCNT and poly (vinylidene fluoride-co-hexafluoropropylene) (PVDF-HFP) as materials of electrode film. MWCNT/PVDF-HFP nanofiber sheet was fabricated by electrospinning. Electrospinning is one of the popular methods for fabricating nanofiber sheet [23], because it can be used in the processing of polymers into nanofiber sheet by applying a high voltage between the grounding collector and spinning nozzle [24]. Nanofibers have outstanding properties such as large surface area, high length/diameter ratio, and flexible surface functionality [25]. Therefore, actuators which are fabricated by electrospinning have several advantages of good dispersion, enhanced flexibility, and ability to produce different form by varying essential factors [26].

In this study, we fabricated MWCNT/PVDF-HFP nanofiber sheet using electrospinning method and adjusted flow rate and polymer concentration in electrospinning solutions for optimum performance in order to obtain nanofiber sheet having less spherical substances called beads, because beads is generated by the incorrect concentration of polymer solution, flow rate, and viscosity of electrospinning solution. We examined morphologies and characterizations of MWCNT/PVDF-HFP nanofiber sheet such as electrical conductivity and mechanical strength. Furthermore, we fabricated carbon nanotube polymer actuator using nanofiber electrode films and operated actuators at DC 6 V.

2. Experimental

2.1. Materials

MWCNTs (purity of 95%, length of 1-10 μm, number of walls are 3-15) were purchased from Plasma Chem. and used as electrical conductive and mechanical strength material. The ionic liquids (IL), 1-Ethyl-3-methylimidazolium tetrafluoroborate (EMI[BF_4]) and PVDF-HFP (M_w = ~400 kDa) were purchased from Aldrich and their chemical structures are shown in Figure 1(a) and (b). 4-Methyl-2-pentanone (MP) ≧99.5%, propylene carbonate anhydrous (PC), and *N, N*-dimethylformamide (DMF) were purchased from Nacalai tesque. The chemicals mentioned above were used without any further purification.

2.2. Fabrication of the electrode film

MWCNTs were dispersed in DMF using ultra sonication for more than 24 h. Then, PVDF-HFP was dissolved in the obtained MWCNTs dispersion at 120°C for 2 h. The weight ratio of MWCNTs to PVDF-HFP was in a range from 0.0 to 10.0wt%. Using this MWCNT/PVDF-HFP solution, MWCNT/PVDF-HFP nanofiber electrode films were fabricated by electrospinning method. Figure 2 shows a schematic illustration of the experimental set-up for electrospinning [26, 27]. The MWCNT/PVDF-HFP solution for electrospinning was injected to a syringe pump and electrospinning was performed at 10 kV to the syringe nozzle. Flow rate of the solution was 0.2-3.0 mL/h. An 18-gauge spinning nozzle with a 0.94 mm inner diameter was employed. The distance between the end of the nozzle and the surface of the rotary collector was 3 cm. The rotary collector for collecting nanofibers was 10 cm diameter, 15 cm width. In order to fabricate aligned MWCNT/PVDF-HFP nanofibers [28-30], the rotary collector was rotated at 1.5-7.3 m/s (300-1400 rpm).

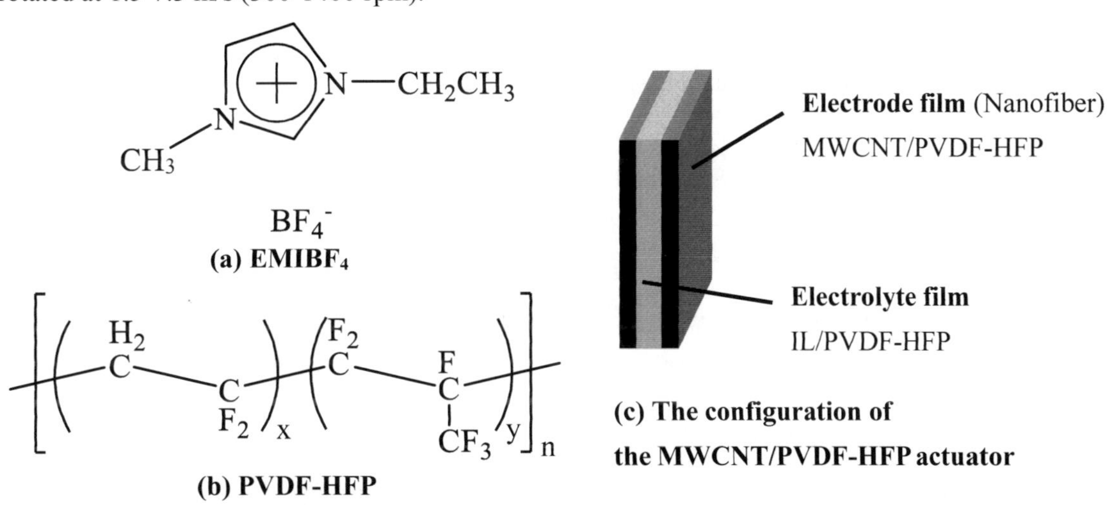

Figure 1. Chemical structures of (a) $EMIBF_4$, (b) PVDF-HFP and (c) configuration of the MWCNT/PVDF-HFP actuator.

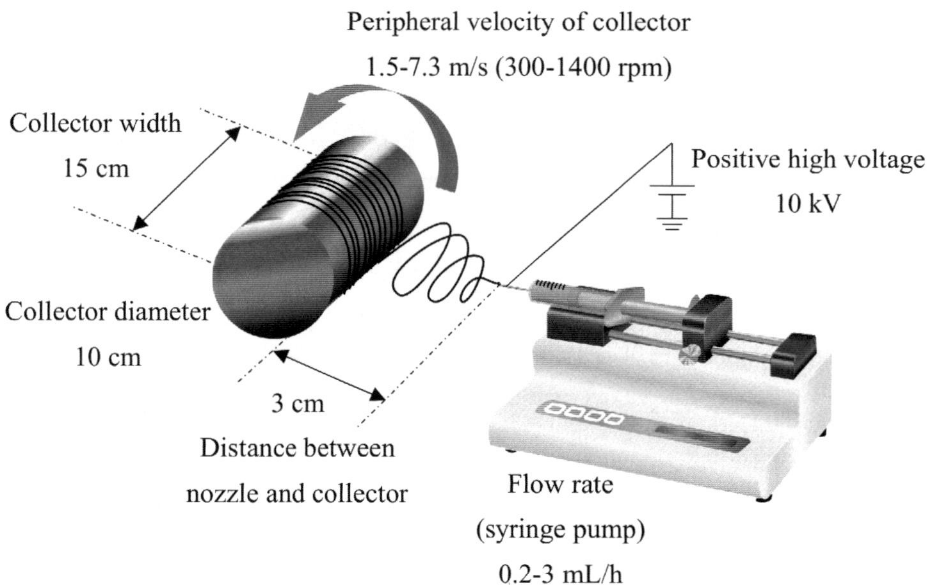

Figure 2. Schematic illustration of the experimental set-up for electrospinning.

2.3. Fabrication of the electrolyte film

The electrolyte film was fabricated by casting 300 μL of electrolyte solution consisting of 80 μL of $EMIBF_4$ (in Figure 1(a)), 130 mg of PVDF-HFP (in Figure 1(b)), 3 mL of MP, and 207 μL of PC on aluminum substrate (the size is 2.5 cm × 2.5 cm). The solvent was then completely removed at 80°C for 2 h. The thickness of the electrolyte film was measured using a digital micrometer (Mitutoyo).

2.4. Fabrication of actuator using MWCNT/PVDF-HFP nanofiber electrode film

The configuration of MWCNT/PVDF-HFP actuator [31] is illustrated in Figure 1(c). An actuator had a three-layer structure having the electrolyte film between the MWCNT/PVDF-HFP nanofiber sheets and is fabricated by hot-press at 130°C for 10 min with 7.5 MPa. The nanofiber electrode film and the electrolyte film, with thickness of 20-40 μm and 50-60 μm, were used to fabricate actuator.

2.5. Characterization

A scanning electron microscope (SEM, KEYENCE VE-8800) was operated at 5-8 kV, and the average nanofiber diameters were measured from SEM images using the software package (Image J) from 100 places/specimen. The alignment intensity of MWCNT/PVDF-HFP nanofibers was measured with SEM images using the software package (Fiber Orientation Analysis Ver. 8.13) [32]. Electrical conductivities of aligned MWCNT/PVDF-HFP nanofibers in the alignment direction were measured by the two-terminal method with a digital multimeter (Model617, TOYO corporation) at room temperature and relative humidity of 40-50%. Mechanical strength of aligned MWCNT/PVDF-HFP nanofibers in the alignment direction were measured with a tensile tester (FGPX-02, NUDEC-SHIMPO) at an elastic stress speed of 10 mm/min at room temperature. The electrical conductivity and the mechanical strength measurement were performed 3 times/specimen. Viscosity of MWCNT/PVDF-HFP solution was measured with VISCOMATE (CBC Co., Ltd. VM-10A-L).

3. Results and discussion

3.1. Morphology of MWCNT/PVDF-HFP nanofiber sheet

3.1.1. Effect of flow rate and polymer concentration

Electrospinning is popular method for fabricating nanofiber. However, this method needs to adjust various condition such as applied voltage, polymer concentration in spinning solution, and flow rate [33, 34]. Therefore, we tried various condition of electrospinning such as flow rate and polymer concentration in order to search for the best condition for fabricating MWCNT/PVDF-HFP nanofiber sheet. Figure 3(a), (b), and (c) show the SEM images of MWCNT/PVDF-HFP nanofiber sheet fabricated under different flow rate. The flow rate for each sample was 3, 1 and 0.2 mL/h, respectively. In this experiment, we fabricated MWCNT/PVDF-HFP nanofiber sheet using solution containing 10wt% CNT and fixed peripheral velocity of collector at 7.3 m/s. In Figure 3, spherical substances called beads decreased as the flow rate decreased. The beads are made in places of nanofiber under incorrect conditions of electrospinning because the nanofiber formation unstabilized due to the instability of the jet of polymer solution [35].

Figure 4(a), (b) and (c) show the SEM images of MWCNT/PVDF-HFP nanofiber sheet fabricated using different polymer concentration of 10, 12 and 15wt% in electrospinning solution, respectively. In this experiment, we fabricated MWCNT/PVDF-HFP nanofiber sheet using solution containing 10wt% CNT. In addition, we fixed peripheral velocity of collector at 7.3 m/s and flow rate at 1mL/h. As shown in Figure 4, the beads decreased as polymer concentration increased. In general, when the concentration of solution increases, the viscosity of solution also increases. Therefore, I investigated the

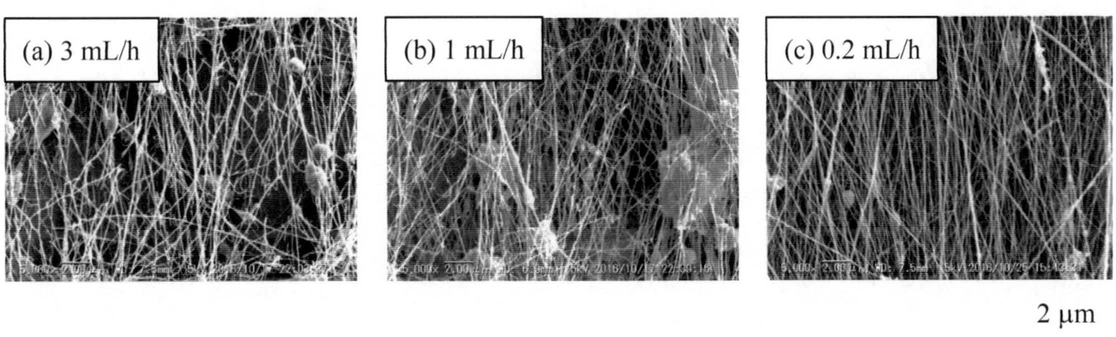

Figure 3. SEM images of MWCNT/PVDF-HFP nanofiber sheet fabricated under different flow rate of (a) 3 mL/h, (b) 1 mL/h and (c) 0.2 mL/h.

IOP Conf. Series: Journal of Physics: Conf. Series **924** (2017) 012005 doi:10.1088/1742-6596/924/1/012005

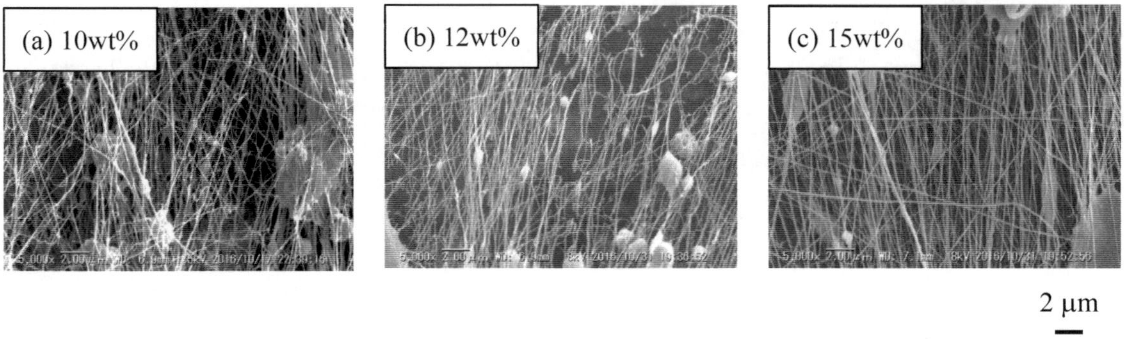

Figure 4. SEM images of MWCNT/PVDF-HFP nanofiber sheet fabricated using different polymer concentration of (a) 10wt%, (b) 12wt% and (c) 15wt% in electrospinning solution.

relationship between the viscosity of the electrospinning solution and nanofiber formation.

Figure 5 shows the viscosity of electrospinning solution as a function of CNT weight ratio for two types of polymer fraction in solution. Figure 6 shows the SEM images of MWCNT/PVDF-HFP nanofiber sheet containing different CNT weight ratio of (a)2wt%, (b)5wt% and (c)10wt% (polymer 15wt% in solution), respectively. In Figure 6, as the CNT weight ratio to polymer increased, the viscosity of electrospinning solution increased. In Figure 6, as the CNT weight ratio increased, it can be confirmed form SEM images that nanofibers with less beads are fabricated. Therefore, we considered that the viscosity of electrospinning solution affected the nanofiber formation. We speculated from Figure 3-6 that it is important for fabricating nanofiber sheet with less beads to adjust flow rate and the viscosity of spinning solution.

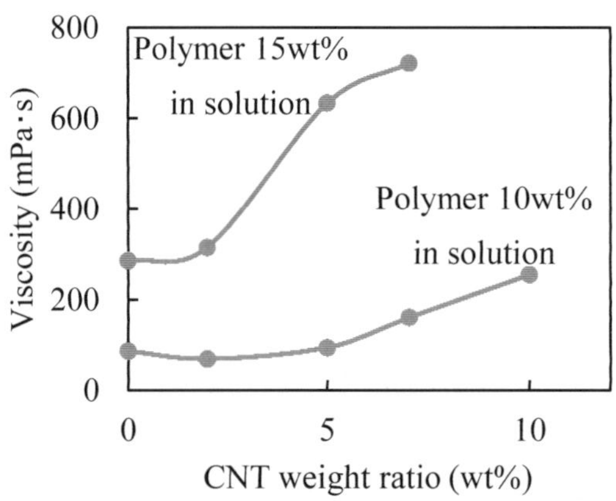

Figure 5. The viscosity of electrospinning solution as a function of CNT weight ratio for two types of polymer weight in solution.

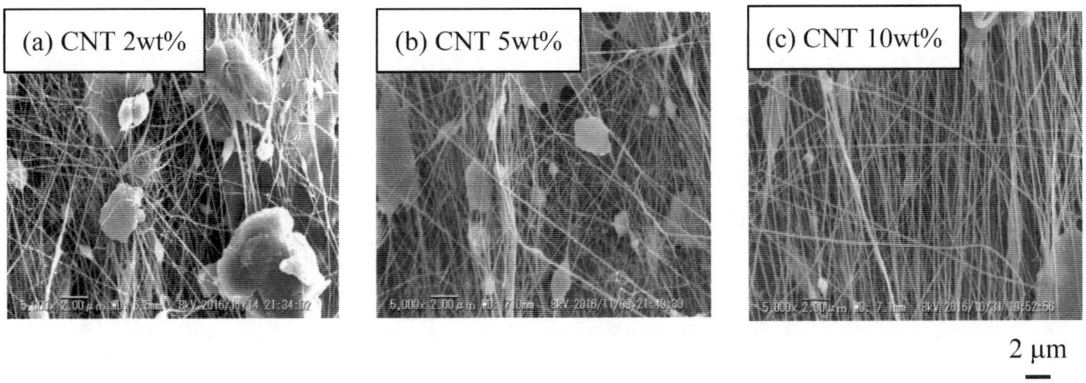

Figure 6. SEM images of MWCNT/PVDF-HFP nanofiber sheet containing different CNT weight ratio of (a) 2wt%, (b) 5wt% and (c) 10wt% (polymer 15wt% in solution).

3.1.2. Effect of peripheral velocity of collector

We also examined the effect of peripheral velocity of collector on nanofiber alignment. Figure 7 shows SEM images of MWCNT/PVDF-HFP nanofiber sheet collected with different peripheral velocity (1.5-7.3 m/s) and Figure 8 shows the alignment intensity [32] of MWCNT/PVDF-HFP nanofiber sheet as a function of peripheral velocity of collector which was measured from SEM images of Figure 7 and other data (not shown). As shown in Figure 7 and 8, the nanofibers were aligned as the peripheral velocity of collector increased. There was report that the trajectory of the continuous electrospinning jet is affected by the rotation of the collector [34]. In electrospinning, when the nanofibers are attached electrostatically onto collector surface, it stretches the following tow of nanofiber from its spiraling path to align with the rotation direction of the collector. Therefore, the effective draw is increased when the peripheral velocity of collector increased, resulting in better alignment of the collected nanofibers. Therefore, we

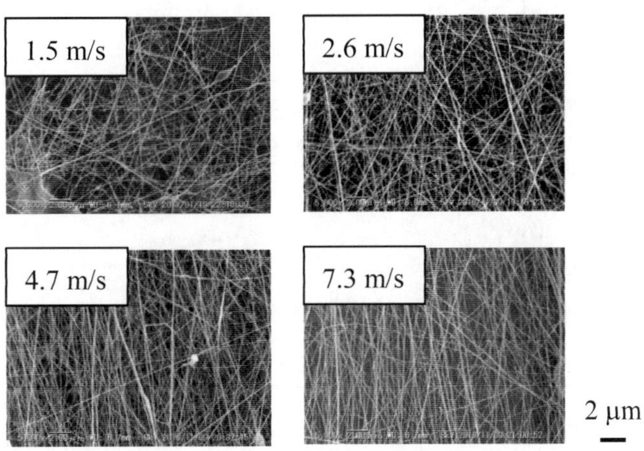

Figure 7. SEM images of MWCNT/PVDF-HFP nanofiber sheet collected different speed (1.5-7.3 m/s).

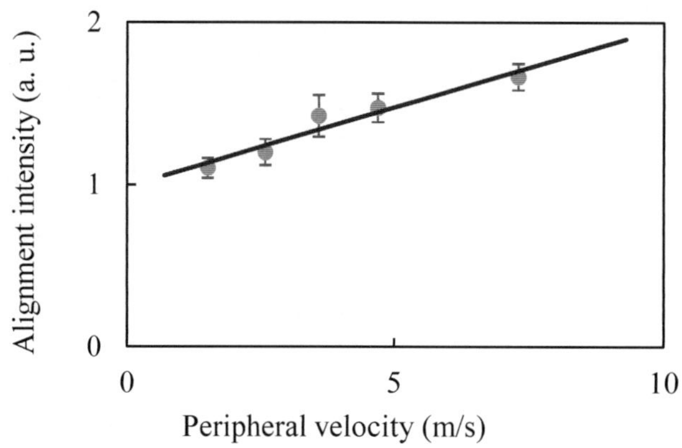

Figure 8. The alignment intensity of MWCNT/PVDF-HFP nanofiber sheet as a function of peripheral velocity of collector.

confirmed that nanofibers fabricated by electrospinning are able to align along rotating direction at high peripheral velocity of collector.

3.1.3 Effect of CNT weight ratio on nanofiber diameter

Figure 9 shows the average diameter of aligned MWCNT/PVDF-HFP nanofibers as a function of CNT weight ratio. In Figure 9, the average diameters of aligned MWCNT/PVDF-HFP nanofibers were 170 (CNT 0wt%), 120 (CNT 2wt%), 110 (CNT 5wt%), 100 (CNT 7wt%) and 90 nm (CNT 10wt%), respectively. The average diameters of aligned MWCNT/PVDF-HFP decreased as the CNT weight ratio increased. As Mazinani et al. have explained [36], we considered that high electrical conductivity of CNT affected the decrease of nanofiber diameter because of increased stretching force in nanofibers by increased charge in sprayed liquid drops from spray nozzle during electrospinning.

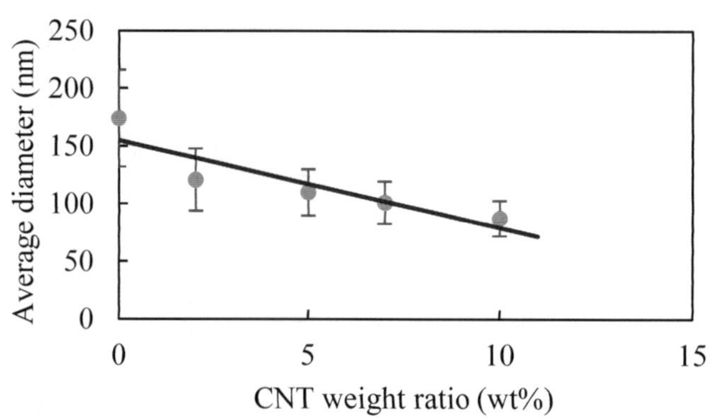

Figure 9. The average diameters of aligned MWCNT/PVDF-HFP nanofibers as a function of CNT weight ratio.

3.2. Characteristics of MWCNT/PVDF-HFP nanofiber sheet

The electrical conductivity and the mechanical strength of aligned MWCNT/PVDF-HFP nanofiber sheet were measured in the alignment direction. Figure 10(a) shows the electrical conductivities of aligned MWCNT/PVDF-HFP nanofiber sheets as a function of CNT weight ratio to the weight of PVDF-HFP in nanofiber sheet. Their electrical conductivities were measured by two-terminal method at room temperature and relative humidity of 40-50%. As shown in Figure 10(a), electrical conductivities of MWCNT/PVDF-HFP nanofiber sheets were 2.9×10^{-10} (CNT 0wt%), 6.6×10^{-2} (CNT 2wt%), 1.3 (CNT 5wt%), 3 (CNT 7wt%), and 12 $S \cdot m^{-1}$ (CNT 10wt%), respectively. The electrical conductivity in the alignment direction increased as the CNT weight ratio increased. These results are considered due to good electrical conductivity of CNTs. In addition, we also considered that increase of electrical conductivity is not linear but curvilinear as the CNT concentration increased, which is because of rapid increase of connection between MWCNTs in PVDF-HFP nanofibers around the 7wt%(CNT weight ratio). Figure 10(b) shows the mechanical strengths of aligned MWCNT/PVDF-HFP nanofiber sheet in the alignment direction as a function of the CNT weight ratio in nanofiber sheets. Their mechanical strengths were measured by measuring tensile strengths using a tensile tester at an elastic stress speed of 10 mm/min at room temperature. In Figure 10(b), tensile strengths of aligned MWCNT/PVDF-HFP nanofiber sheet were improved by introducing CNT in nanofiber due to good mechanical strength of CNTs.

3.3. Operating of actuator

In order to confirm whether operating MWCNT/PVDF-HFP nanofiber electrode actuator is possible, we tried to fabricate polymer actuator using MWCNT/PVDF-HFP nanofiber sheets for electrodes, which sandwiched IL/PVDF-HFP electrolyte film by hot-press. We used MWCNT/PVDF-HFP

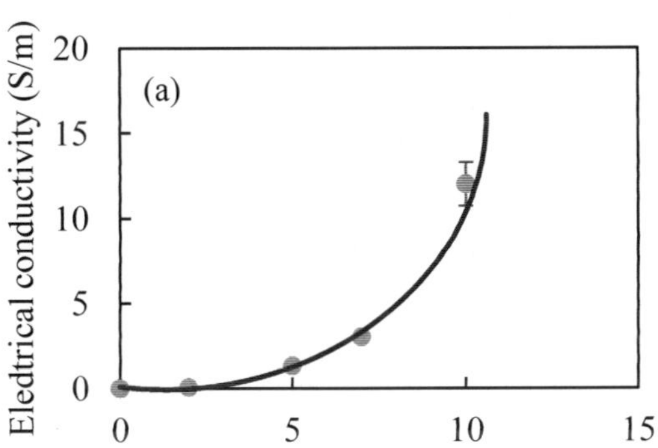

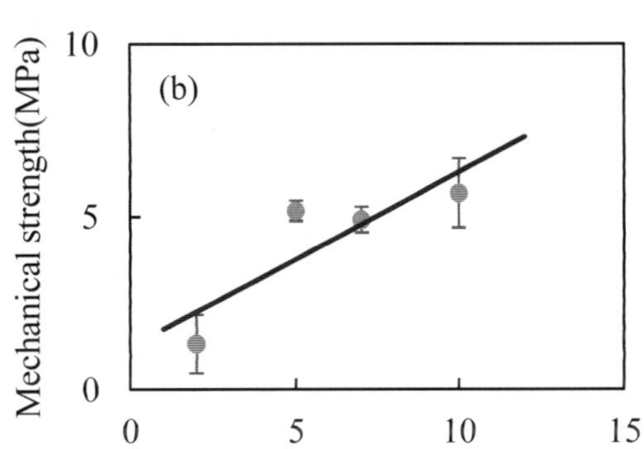

Figure 10. (a)The electrical conductivity and (b) the mechanical strength of aligned MWCNT/PVDF-HFP nanofiber sheet in the alignment direction as a function of CNT weight ratio.

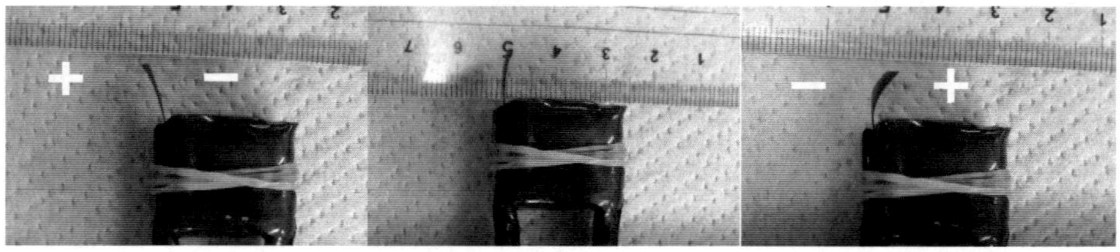

(a) - 6 V (b) 0 V (c) 6 V

Figure 11. The optical images of bending actuator at (a) - 6 V, (b) 0 V and (c) 6 V.

nanofiber sheet same as experimental condition in Figure 3 (b). Figure 11 shows the optical images of bending actuator at (a) - 6 V, (b) 0 V and (c) 6 V. In Figure 11, we confirmed that actuator was bending. This actuator bent to the left at (a) - 6 V and to the right at (c) 6 V.

4. Conclusions

In this paper, we fabricated carbon nanotube polymer actuator consisted of a $EMIBF_4$/PVDF-HFP electrolyte film between MWCNT/PVDF-HFP nanofiber electrode films fabricated by electrospinning method.

Nanofibers with less beads are fabricated by adjusting flow rate and polymer concentration. In addition, it was found that the nanofibers were aligned as the peripheral velocity of collector increased, electrical conductivity and mechanical strength of nanofibers increased as the MWCNT weight ratio to the polymer increased, and the actuator moved to the right or left at DC 6 V between the electrodes.

It was considered above results that EAP actuator has wide range of applications such as artificial muscle and so on because of their various advantage such as flexibility and lightness. Therefore, it is important to improve the EAP actuator performance, and this study has possibility of opening a novel route for improving EAP actuator performance using polymer nanofibers containing CNT.

References

[1] Terasawa N, Ono N, Mukai K, Koga T, Higashi N and Asaka K 2012 *Carbon* **50** 311-20

[2] Mukai K, Asaka K, Kiyohara K, Sugino T, Takeuchi I, Fukushima T and Aida T 2008 *Electrochimica Acta* **53** 5555-62

[3] Zhiwei Z, Qingwei L, Luzhuo C, Changhong L and Shoushan F 2016 *J. Mater. Chem. B* **4** 1228-34

[4] Takeuchi I, Asaka K, Kiyohara K, Sugino T, Terasawa N, Mukai K, Fukushima T and Aida T 2009 *Electrochimica Acta* **54** 1762-68

[5] Baughman R H *et al.* 1999 *Science* **284** 1340-44

[6] Terasawa N, Ono N, Mukai k, Koga T, Higashi N and Asaka K 2012 *Sensors and Actuators B* **163** 20-8
[7] Kaneto K 2016 *Journal of Physics: Conference Series* **704** 012004
[8] Joanne Y, Feng D, Kit-Lun Y, Chun-Wah M Y, Ting-Ting L and Wing-Hang C 2012 *Sensors and Actuators B* **162** 76-81
[9] Hara S, Zama T, Takashima W and Kaneto K 2004 *Polymer Journal* **36** 151-61
[10] Otero T F 2009 *J. Mater. Chem.* **19** 681-9
[11] Jie D, Dezhi Z, Geoffrey S, Gordon W, Stewart F, Maria F and Douglas M 2003 *Chem. Mater.* **15** 2392-98
[12] Mark P S, Mustafizur R and Christopher S B 2003 *European Polymer Journal* **39** 1947-53
[13] Fukumoto K, Yoshizawa M and Ohno H 2005 *J. Am. Chem. Soc.* **127** 2398-99
[14] Inpil K *et al.* 2006 *Composites Part B:Engineering* **37** 382-94
[15] Fukushima T, Asaka K, Kosaka A and Aida T 2005 *Angew. Chem.* **117** 2462-65
[16] Fukushima T, Kosaka A, Ishimura Y, Yamamoto T, Takigawa T, Ishii N and Aida T 2003 *Science* **300** 2072-74
[17] Ebbesen T W, Lezec H J, Hiura H, Bennett J W, Ghaemi H F and Thio T 1996 *Nature* **382** 54-6
[18] Shiraishi S, Kurihara H, Okabe K, Hulicova D and Oya A 2002 *Electrochemistry Communications* **4** 593-98
[19] Treacy M M J, Ebbesen T W and Gibson J M 1996 *Nature* **381** 678-80
[20] Bahghman R H, Zakhidov A Z and Heer W A 2002 *Science* **297** 787-92
[21] Terasawa N and Asaka K 2017 *Sensors and Actuators B* **240** 536-42
[22] Asai H, Kawai T, Shimada N, Sakai T and Nakane K 2015 *Sensors and Actuators B* **214** 76-81
[23] Narinder K, Vipin K and Sanjay R D 2016 *SpringerPlus* **5** 483
[24] Huang Z M, Zhang Y Z, Kotani M and Ramakrishna S 2003 *Composites Science and Technology* **63** 2223-53
[25] Shu-Hua W, Yong W, Bin S, Ling-Zhi L and Weijiang X 2014 *Nanoscale Research Lettters* **9** 522
[26] Joung-Man P, Ga-Young G, Zuo-Jia W, Dong-Jun K, Pyeong-Su S, Jin-Yeong C and DeVries K L 2015 *Advanced Composite Materials* **25** 305-16
[27] Minjae K, Yeongseon K, Kyung M L, Seon Y J, Eunsoo L, Sung H B and Sang E S 2016 *Carbon* **99** 607-18
[28] Edwards M D, Mitchell G R, Mohan S D and Olley R H 2010 *European Polymer Journal* **46** 1175-83
[29] Luana P, Canan D, Yewang S, Yihui Z, Salvatore G, Dario P, Yonggang H and John A R 2013 *Nature Communications* **4** 1633
[30] Mokhatari F, Latifi M and Shamshirsaz M 2015 *The Journal of Textile Institute* **107** 1037-55
[31] Terasawa N and Takeuchi I 2010 *Sensors and Actuators B* **145** 775-80

[32] Enomae T, Han Y H and Isogai A 2006 *Nordic Pulp and Paper Research Journal* **21** 253-59
[33] Jianfen Z, Aihua H, Junxing L and Charles C H 2007 *Macromol. Rapid Commun.* **28** 2159-62
[34] Fennessey S F and Farris R J 2004 *Polymer* **45** 4217-25
[35] Fong H, Chun I and Reneker D H 1999 *Polymer* **40** 4585-92
[36] S Mazinani, A Ajji and C Dubois 2009 *Polymer* **50** 3329-42

IOP Conf. Series: Journal of Physics: Conf. Series **924** (2017) 012006 doi:10.1088/1742-6596/924/1/012006

Photophysical Characterization and BSA Interaction of Direct Ring Carboxy Functionalized Symmetrical squaraine Dyes

Maryala Saikiran[1], Shyam S Pandey[1] Shuzi Hayase[1] and Tamaki Kato[1]

[1] Graduate School of Life Science and Systems Engineering, Kyushu Institute of Technology, 2-4 Hibikino, Wakamatsu, Kitakyushu 808-0196, Japan

e-mail:msaikiran87@gmail.com / shyam@life.kyutech.ac.jp

Abstract: A series of far-red sensitive symmetrical squaraine dyes bearing direct –COOH functionalized indole ring were synthesized, characterized and subjected to photophysical investigations. These symmetrical squaraine dyes were then subjected to investigate their interaction with bovine serum albumin (BSA) in Phosphate buffer solutions. All the squaraine dyes under investigation exhibit intense and sharp optical absorption mainly in the far-red wavelength region from 550 nm -700 nm having very high molar extinction coefficients from 1.3 X 10^5 $dm^3.mol^{-1}.cm^{-1}$. A very small Stokes shift of 10-17 nm indicates the rigid conformational structure of squaraine chromophore. Interaction of these dyes with BSA leads to not only enhanced emission intensity but also bathochromically shifted absorption maximum due to formation of dye-BSA conjugate. These dyes bind strongly with BSA having about an order of magnitude higher binding constant as compared to the reported squaraine dyes. Amongst the symmetrical squaraine dyes investigated in this work one bearing substituents like trifluorobutyl as alkyl chain at N-position of indole ring and carboxylic acid on benzene ring at the terminal **(SQ-26)** exhibited highest association with the BSA having very high binding constant 8.01 x 10^6 M^{-1}.

1. Introduction

Biomedical diagnosis is a fast growing area in order to afford high quality of life by early disease diagnosis. In recent years there has been dramatic development towards biosensors due to their applications in chemical analysis and medical diagnosis. Optical imaging has emerged as most common imaging practice in medical research due to its high sensitivity, low cost and non-invasive nature [1]. Near infrared (NIR) fluorescence probes have their potential attraction towards non-invasive *in vivo* imaging due to their highly diminished auto-fluorescence from the biological fluids and deep tissue penetration [2]. NIR probes have drawn attention in several areas of the biomedical field, including tissue perfusion, vascular mapping, inflammation monitoring and tumor diagnosis. The squaraine dyes belongs to the interesting organic chromophoric system, where oxocyclobutenolate core is linked by aromatic or heterocyclic components at both ends leading to donor–acceptor–donor molecular motif. Sensitizing dyes belonging to squaraine family are of considerable interest as fluorescent labels/probes for biological and pharmaceutical research owing to their tunable absorption/emission from visible to NIR wavelength region, high molar absorptivity and reasonably good quantum yields [3-4]. By logical molecular design, the absorption and emission behaviour of these dyes can be tuned from visible to IR wavelength making them suitable for fluorescence imaging since they are exterior to the self-absorption and auto-fluorescence regions of biological matrices [5]. The unique chemical and physical properties of squaraine dyes find their versatile applications in the area as NIR fluorescent probes, environmental sensors [6], molecular sensors [7-8], bio imaging and biochemical labelling [9-10] which have gathered huge attentions from the scientific community.

The present work deals with the synthesis of four direct carboxy ring functionalized symmetrical NIR squaraine dyes (**SQ-3, SQ-26, SQ-56 and SQ-57**) from their respective quaternary heterocyclic ammonium salt containing N-alkylated group. The presence of direct ring functionalized carboxy groups provide the capability of covalent coupling with biomolecules such as peptides, oligonucleotides and proteins to enhance their application potential as florescence probes. In order to explore their potential application as fluorescent probes to sense protein in phosphate buffer solution (PBS), these dyes were subjected to investigate their interaction using Bovine serum albumin (BSA) as a model protein to observe the influence of alkyl chain with varying functional groups of the dyes.

2. Materials and Methods

1-Iodobutane, 1,1,1-trifluoro-4-iodobutane, 1,3-propanesultone, Ethyl-4-bromobutanoate used in the present synthesis were purchased from Tokyo Kasei Co. Ltd. Solvents (reagent grade, Wako chemical company) and Squaric acid was purchased from Alfa Aesar and used as received. Synthesized synthesized squaraine dyes and dye intermediates were analysed by high performance liquid chromatography (HITACHI) equipped with chromolith analytical column (RP – 18e, φ 4.6 mm × 100 mm) using methanol-water solvent gradient. Fig. 1 indicates the purity of final symmetrical squaraine dyes used for present investigation which were verified by HPLC. Mass of the intermediates as well as final SQ dyes were confirmed by MALDI-TOF-mass or fast ion bombardment (FAB) mass in positive ion monitoring mode. For final symmetrical squaraine dyes, high resolution FAB-mass (HR-MS) in positive ion monitoring mode was also measured. Nuclear Magnetic Resonance (NMR) spectra were recorded on a JEOL JNM A500 MHz spectrometer in CDCl3 or d6-DMSO with reference to TMS for structural elucidation.

2.1 BSA Interactions

The protein (BSA)-dye interactions were conducted using PBS (0.1 M at pH 7.4) and squaraine dye solutions (2μM), prepared by the addition of 100 μl of 0.1 M dye solution in DMF to the various concentrations of PBS/BSA solutions in the concentration range of (0-10 μM). The final solutions were stirred at room temperature before recording the respective electronic absorption and fluorescence emission spectra. To compare the interactions between protein and dye, the apparent binding constants (K_a) were also calculated from the following titrations of dyes at 25^{0}C. Considering BSA/dye association in 1:1 ratio, the constant (K_a) was calculated using the equation 1 [11].

$$\frac{1}{(F_x - F_0)} = \frac{1}{(F_\infty - F_0)} + \frac{1}{K_a[BSA]}\frac{1}{(F_\infty - F_0)} \quad (1)$$

Where, F_0, F_X, and F_∞ are the fluorescence intensities of dyes in the absence, presence of BSA and at a concentration of complete interaction, respectively, while [BSA] is the protein concentration. Equation [1] can be modified as

$$\frac{(\mathrm{F}_\infty - \mathrm{F}_0)}{(\mathrm{F}_\mathrm{x} - \mathrm{F}_0)} = 1 + \frac{1}{\mathrm{K}_\mathrm{a}[\mathrm{BSA}]} \quad (2)$$

IOP Conf. Series: Journal of Physics: Conf. Series **924** (2017) 012006 doi:10.1088/1742-6596/924/1/012006

The binding/association constant (K_a) values for the interaction between the BSA and Squaraine dyes were calculated from the slopes of the corresponding plots between the $(F_\infty - F_0)/(F_X - F_0)$ as a function of $[BSA]^{-1}$ as per the equation (2).

2.2 Synthesis of SQ dyes and dye intermediates

Aromatic ring carboxy functionalized indole 2,3,3-trimethyl-3H-indole-5-carboxylic acid was synthesized following the method reported [12]. Symmetrical squaraine dyes and dye intermediates of 5-carboxy-2,3,3-trimethyl-indole have been synthesized following the method as shown in Scheme 1.

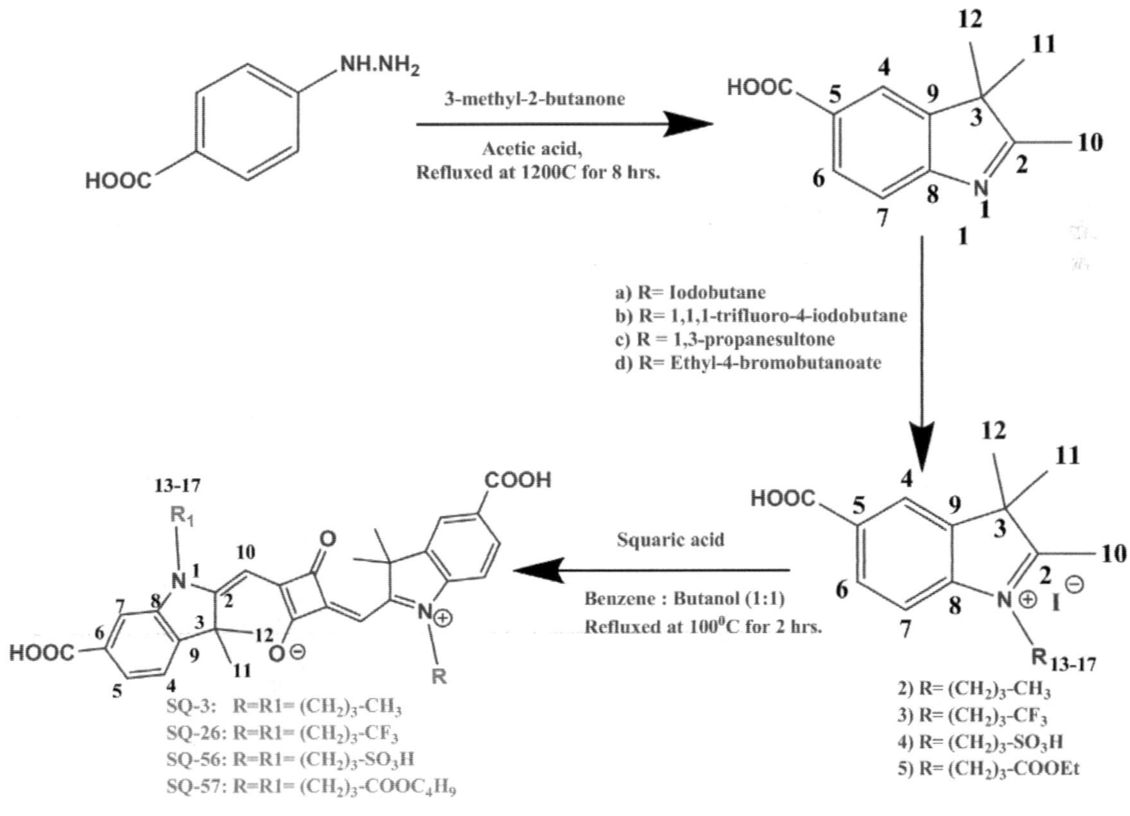

Scheme 1: Synthesis of symmetrical squaraine dyes.

2.2.1. Synthesis of 2,3,3-trimethyl-3H-indole-5-carboxylic acid (1)

In a round bottom flask fixed with condenser, 4-hydrazinobenzoic acid (5.0 g; 32.85 mmol), glacial acetic acid (80 ml), sodium acetate (5.5 g; 67 mmol) and 3-methyl-2-butanone (4.45 g; 51.5 mmol) were added. Reaction mixture was refluxed at 120^{0}C for 8 h leading to brown suspension. Upon the completion of reaction, acetic acid was evaporated followed by addition of 9:1 water methanol mixture on ice-bath leading to precipitation. Residue was filtered and dried giving 3.7 g of titled compound as off white powder in 56% yield. HPLC analysis of product suggests that compound was 100% pure. FAB-mass (measured 203; calculated 203.09) and ^{1}H NMR (500 MHz, CDCl3): 7.99 (s,

H-4), 7.93 (d, J = 8.0 Hz, H-6), 7.59 (d, J = 8.0 Hz, H-7), 2.26 (s, 3H, H-10), 1.28 (s, 6H, H11, 12) confirms the successful synthesis of the compound.

2.2.2. Synthesis of 5-carboxy-2,3,3-trimethyl-1-alkyl-3H-indolium iodide (2-5)

2,3,3-trimethyl-3H-indole-5-carboxylic acid (1, 1 equiv.) and (a, 3 equiv, in Acetonitrile, 95^{0}C), (b, 3 equiv, in propionitrile, 100^{0}C), (c, 3 equiv, in DCB, 135^{0}C), (d, 3 equiv, in propionitrile, 100^{0}C), were added and reaction mixture was refluxed for (a: 48 h, b: 24 h, c: 72 h, d: 24 h) under nitrogen atmosphere to give corresponding 5-carboxy-N-alkyl-indoium iodides (2-5). After completion of the reaction as monitored by HPLC, solvent was evaporated and the crude product was washed with ample diethyl ether giving the titled compound. The physical and spectral data of N-alkyl-indolium iodides (2-5) are as follows.

2.2.3. Synthesis of 1-butyl-5-carboxy-2,3,3-trimethyl-3H-indolium (2): Yield 77% with 97% purity as confirmed by HPLC. FAB-mass (measured 260.0; calculated 260.16) confirms successful synthesis of the compound [13].

2.2.4. Synthesis of 5-Carboxy-2,3,3-trimethyl-1-Trifluorobutyl-3H-indolium iodide (3): Yield 46% with 99% purity as confirmed by HPLC. FAB-mass (measured 441.0; calculated 441.04) confirms the successful synthesis of the compound [14].

2.2.5. Synthesis of 5-carboxy-2,3,3-trimethyl-1-(3-sulfopropyl)-3H-indol-1-ium (4): Yield 59% with 99% purity as confirmed by HPLC. FAB-mass (measured 326.1066; calculated 326.1057) confirms the successful synthesis of the compound [15].

2.2.6. Synthesis of 5-carboxy-1-(3-ethoxy-3-oxopropyl)-2,3,3-trimethyl-3H-indol-1-ium (5): Yield 59% with 99% purity as confirmed by HPLC. FAB-mass (measured 318.1698; calculated 318.1700) confirms the successful synthesis of the compound [16].

2.2.7. Synthesis of symmetrical squaraine dyes

Symmetrical SQ Dyes (SQ-3, 26, 56 and 57) were synthesized using corresponding carboxy functionalized trimethyl-indolium iodide salt 2-5 ((2 equiv.) and squaric acid (1 equiv.)) in 1-butanol: benzene mixture. Reaction mixture was refluxed for 18 h using Dean–Stark trap for azeotropic removal of water. After completion of reaction, reaction mixture was cooled, solvent was evaporated and product was purified by silica gel column chromatography using chloroform: methanol as eluting solvent. The physical and spectroscopic data of symmetrical SQ dyes are as follows;

***2.2.8. Synthesis of N-butyl substituted squaraine dye SQ-3*:** Yield 64% with HPLC purity 100%. MALDI-TOF-mass (calculated 596.29 and observed 597.25 [M+H]$^+$). HR-MS (calculated 596.288 and observed 596.296 [M]$^+$). 1H NMR (500 MHz, d6-DMSO): d/ppm = 12.85 (b, –COOH), 8.04 (dd, H-6), 7.96 (dd, H-4), 7.43 (dd, H-7), 5.90 (s, H-10), 4.13 (q, 2H, H-13), 1.76 (m, 2H, H-14), 1.70 (s, 6H, H11 + 12), 1.40 (m, 2H, H-15), 0.95 (t, 3H, H-16) confirms the successful synthesis of the dye.

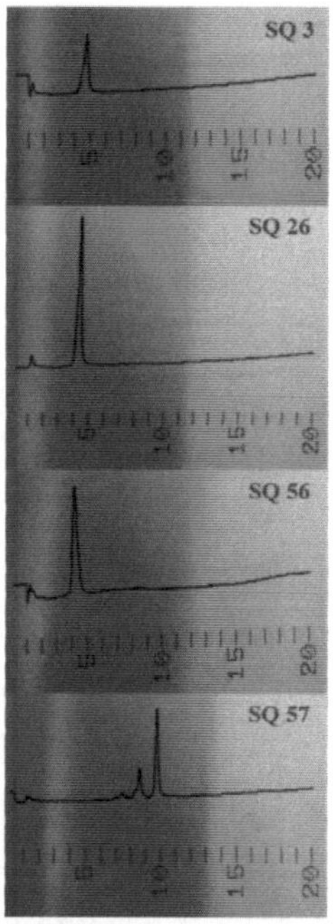

Figure. 1 HPLC chromatogram of symmetrical Squaraine dyes

***2.2.9. Synthesis of N-Trifluorobutyl substituted squaraine dye SQ-26*:** Yield 71% with HPLC purity 100%. FAB-mass (calculated 704.23 and observed 705.0 $[M+H]^+$) confirms the identity of the compound.

***2.2.10. Synthesis of N-sulfopropyl substituted squaraine dye SQ-56*:** Yield 42% with HPLC purity 100%. FAB-mass (calculated 728.17 and observed 729.00 $[M+H]^+$) confirms the identity of the compound.

***2.3.11. Synthesis of N-butoxybutyl substituted squaraine dye SQ-57*:** Yield 42% with HPLC purity > 80%. FAB-mass (calculated 769.9040 and observed 769 $[M]^+$) confirms the identity of the compound.

3. Results and Discussion

3.1. *Photophysical Characterization*

After the successful synthesis and purification, these far-red sensitive dyes were subjected to photophysical investigations pertaining to the electronic absorption and fluorescence emission spectroscopy. The results thus obtained have been summarized in the table 1. Fig. 2 exhibits the solution state electronic absorption and fluorescence emission spectra of symmetrical squaraine dyes in the dimethylformamide (DMF) solution.

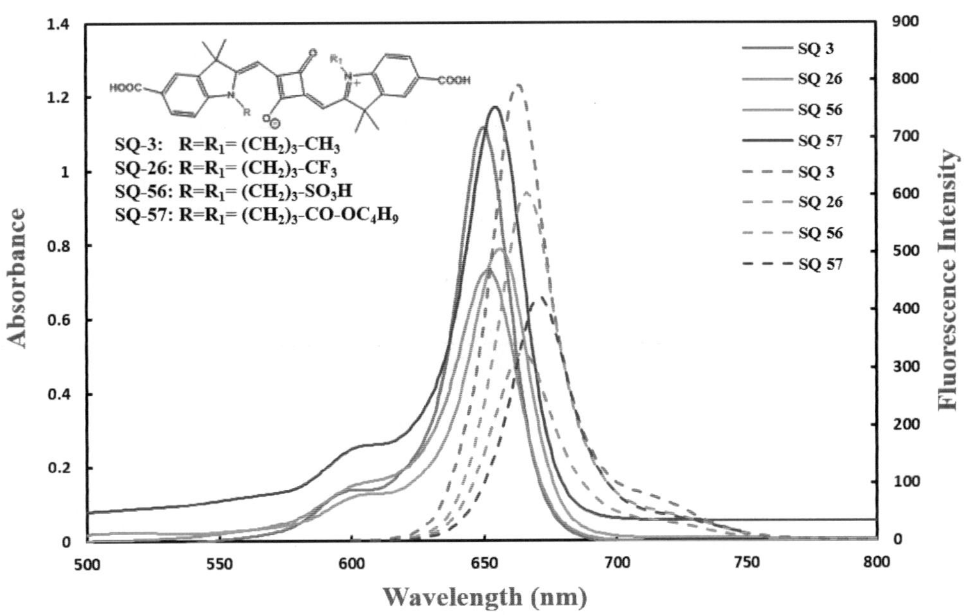

Figure. 2 Electronic absorption (solid line) and fluorescence emission (dashed line) spectra of dyes in DMF solution (5 μM). The inset shows the molecular structure of squaraine dyes.

It can be observed that position of absorption maxima (λ_{max}) in the electronic absorption and fluorescence emission spectra are slightly affected by the different substituents on indole ring. The λ_{max} of symmetrical squaraine dyes ranges from 650 to 656 nm with high molar extinction coefficient ranging from 2.34×10^5 to 1.45×10^5 dm^3 M^{-1} cm^{-1}. The intense optical absorption of these dyes are associated with π–π* electronic transitions. The fluorescence emission spectra were measured slightly below (about 12 to 16 nm) to the resultant λ_{max} of absorption spectra. The dyes exhibit main fluorescence emission band where the peak maxima varies from 662 to 672 nm with a small stoke shift of 10 to 17 nm. This small Stokes shift represents the rigidity of the molecules without having any conformational changes after the photoexcitation.

Table 1. Photophysical properties of squaraine dyes in DMF solution

Sensitizing Dyes	λ (max) Absorption	λ (max) Emission	Stoke Shift	(ε) (dm³ M⁻¹ cm⁻¹)
SQ-3	650 nm	662 nm	12 nm	2.2×105
SQ-26	652 nm	664 nm	12 nm	1.4×105
SQ-56	656 nm	666 nm	10 nm	1.5×105
SQ-57	655 nm	672 nm	17 nm	2.3×105

3.2. Interaction of dyes with BSA

To investigate the interactions between dyes and proteins, PBS has been used as a choice of buffer, which is most widely used in imaging applications for protein folding and ligand binding mechanism [17]. Non-covalent labelling of biomolecules has caught much attention to investigate the interactions between dye and protein molecules. Therefore, in order to avoid the purification steps and chemical reactions of biomolecules with dyes towards the application of optical imaging, non-covalent labelling has been employed [18]. Frank Welder et al were studied the noncovalent labelling of both Bovine Serum Albumin (BSA) and Human Serum Albumin (HSA) with NIR Squarylium dyes using absorption detection techniques [19]. Bovine serum albumin (BSA) is a globular protein which has been prominently used as protein model to investigate the interactions between dye and protein owing to its high homology with HSA in the amino acidic sequences [20]. Therefore, dyes used in this work have also been subjected to investigate their interaction with BSA as a model protein.

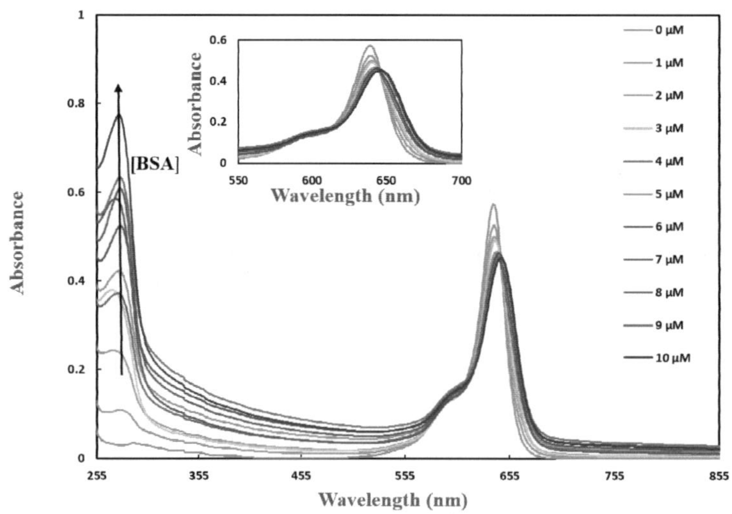

Figure. 3 Electronic absorption spectra of SQ-3 in 0.1 M PBS at different concentrations of BSA for fixed dye concentration of 2 μM.

Fig. 3 depicts the absorption spectra of one of the representative squaraine dyes (SQ-3) in the presence and absence of BSA. It can be seen that there are two different sets of prominent electronic absorption bands in the wavelength region of 250 nm - 300 nm and 620 nm to 680 nm associated with the absorption of BSA and dye SQ-3, respectively. Increase in the BSA concentration led to the gradual increase in intensity of absorption between 250 nm - 300 nm, which was associated with electronic absorption of the protein BSA. However, the interaction of BSA with SQ-3 exhibits slight red-shift from 639 nm to 649 nm upon increasing concentration of BSA. There was initial decrease (hypochromism) in absorption intensity which continues up to 4-6 μM along with the presence of isobestic point and then remains constant upon the further addition of BSA. The isobestic point indicates that the equilibrium between the free and bound protein is simple in all the dyes under investigation. This observed red shift and hypochromicity could be attributed to the presence of hydrophobic environment provided by BSA to the SQ-3 which is an indication for the formation of squaraine dye-BSA conjugate. It is well known that squaraine dyes are prone to dye aggregate formation in the solution owing to their flat molecular structure. Upon interaction with BSA first of all there is dye-BSA conjugate formation leading to disruption of dye aggregates resulting in to red-shifted absorption maximum. At the same time, compaction of the effective π-electron density decreases due to dye-BSA conjugate formation leading to the hypochromicity. Once conjugate formation is finished, further increase in the BSA (> 6 μM) does not affect the decrease in the intensity of SQ-dye rather than only red-shift. This is attributed to the fact that now added additional BSA is involved mainly in the prevention of dye aggregation rather than conjugate formation. Similar behaviour of interaction between squaraine dyes with HSA has also been noticed by Jisha et al during their investigation [21]. It is worth to mention that similar behaviour has also been observed for the other squaraine dyes such as **(SQ-26, SQ-56, and SQ-57)** but SQ-56 shows rather enhanced side vibronic shoulder associated with the dye-aggregation owing to its strong hydrophilic nature due to presence of two highly hydrophobic -SO_3H groups in the side chain.

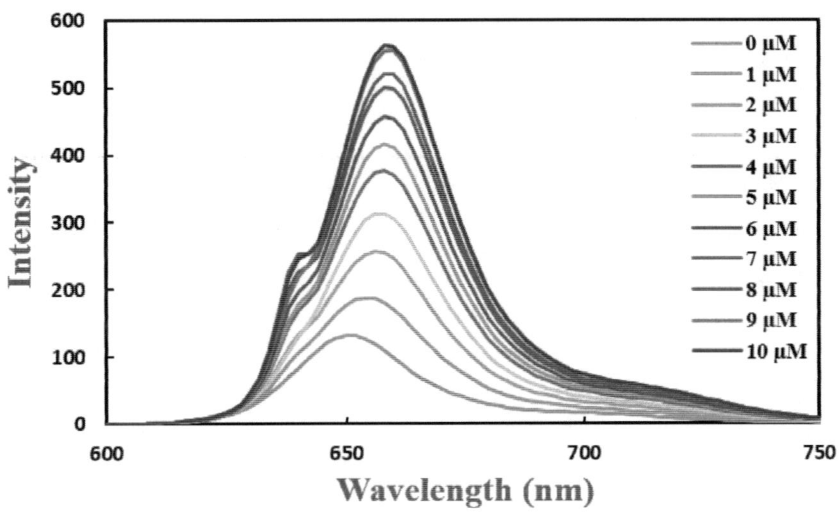

Figure. 4 Fluorescence emission spectra of SQ-3 with varying concentrations of BSA for a fixed dye concentration of 2 μM.

Changes in the fluorescence behaviour of dyes on interactions with biomolecules forms the fundamental basis for fluorescence bio-imaging. Spectroscopic behaviour of squaraine dyes have been reported to be highly sensitive to the environmental conditions due to their self-aggregation [22]. In common, fluorophores bind with protein covalently or non-covalently and non-covalent binding or

IOP Conf. Series: Journal of Physics: Conf. Series **924** (2017) 012006 doi:10.1088/1742-6596/924/1/012006

labelling shows preference for hydrophobic binding sites, where it finds potential application in biomedical science in order to investigate the changes in the conformation of proteins and their binding with drugs [18]. The fluorescence emission spectra of SQ-3 in the presence and absence of BSA is shown in the Fig. 4. It can be seen from Fig. 4 that SQ-3 exhibits an increase in fluorescence intensity along with slight red shift of peak maxima upon the addition of increasing amounts of BSA. The increase in fluorescence intensity could be attributed to the possible noncovalent interactions between the dye and BSA. The bathochromic shift of 10 nm with SQ-3 depicts that BSA microenvironments are less polar than that of PBS due to presence of hydrophobic groups on surface and interiors of BSA. It can be argued that why there was continues increase in the fluorescence intensity of dye upon the addition of BSA where absorption maximum saturates up to 6 μM of BSA concentration. As discussed earlier, BSA is not only involved in the dye-BSA conjugate formation but also prevention of dye aggregation. It has also been reported that there is quenching in the fluoresecence upon the dye aggregate formation [23]. This is the reason why addition of the BSA with concentration > 6 μM leads to further enhancement in the fluorescence intensity of dye by preventing the dye aggregation. Similar results were also observed with the other dyes (**SQ-26, SQ-56 and SQ-57**) with relatively quenched fluorescence intensity for SQ-56. This quenching in the fluorescence emission for SQ-56 could be associated with strong dye aggregation providing additional pathway for non-radiative decay.

BSA concentration dependence of the fluorescence intensity for various symmetrical Squaraine dyes have been shown in the Fig. 5. A linear correlation between fluorescent intensity of dyes as a function of BSA concentration was observed up to 6 μM concentrations of BSA and attained saturation. Though, these dyes exhibit interactions on further addition of BSA, the emission intensity seems to follow nonlinearity, where the change in fluorescence intensity is comparatively less as it attained saturation at 6 μM of BSA. This increase in fluorescent intensity along with red shift of λ_{max} can be attributed to the noncovalent interaction between dyes and protein for the formation of BSA-dye conjugates. It has been widely accepted that drugs or probes interact differently with the protein under consideration and their ability to bind depends on protein concentration [24]. In order to mark the successful application of dyes as a probe, it is necessary that there should not be any damage to the structure of the protein on interaction with dyes, which basically governs the protein function and activity. The conformational changes in protein may occur at the point where probe is located. As the dye concentration increases, there may be more damage in its activity. Therefore, a very low dye concentration of (2 μM) has been used for investigation of the dye-BSA interaction with all the dyes under investigation.

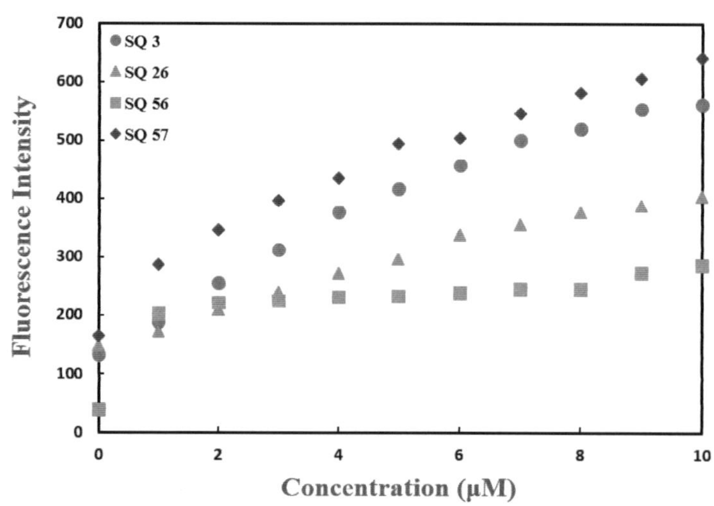

Figure. 5 Plot of florescence emission intensity at peak maxima for Squaraine dyes as a function of BSA concentration. Dye concentration was constant (2 μM) for each of the dyes.

To compare the ability of binding and relative association of dyes with BSA quantitatively, apparent binding constant (K_a) was calculated using equation (2) by plotting ($F_\infty - F_0$)/($F_X - F_0$) as function of inverse of BSA concentration as shown in the Fig. 6. The value of (K_a) was calculated from the slopes of the curve which was found to be 6.2 x 10^6 M^{-1}, 8.0 x 10^6 M^{-1}, 0.3 x 10^6 M^{-1} and 3.9 x 10^6 M^{-1} with the SQ- 3, SQ-26, SQ-56 and SQ-57, respectively. Therefore, symmetrical squaraine dye bearing trifluorobutyl as alkyl chain substituent at `N` position of the indole ring exhibits highest binding affinity and estimated (K_a) is about an order of the magnitude higher as compared to that reported squaraine dyes [25]. Relatively moderate hydrophobicity/hydrophilicity of the SQ-26 is responsible to show the high binding ability to the protein and hence high binding constant. The dye SQ-56 exhibits least binding ability towards BSA, the enhanced hydrophilicity or least hydrophobicity along with pronounced dye aggregate formation might be responsible for the least interaction with protein. BSA possess distinct hydrophobic and hydrophilic active sites site I and Site-II [26]. Upon interaction with the ligands (drug or dyes) the relative hydrophobicity of ligands plays an important role in controlling their interactions with the BSA specifically with Site-I. A perusal of the molecular structures of the dyes under investigation reveals that all of dyes possess nearly same main π-framework, and –COOH functional group apart from the varying alkyl substituents on N-position of the indole unit. The relative hydrophobicity of the molecules are in the order **SQ-57 > SQ-3 > SQ-26 > SQ-56** as confirmed by the HPLC retention time shown in the Fig. 1. Since other structural factors are constant balanced hydrophilic and hydrophobic interactions with Site-I and Site-II of the BSA could be attributed to its strongest affinity with BSA.

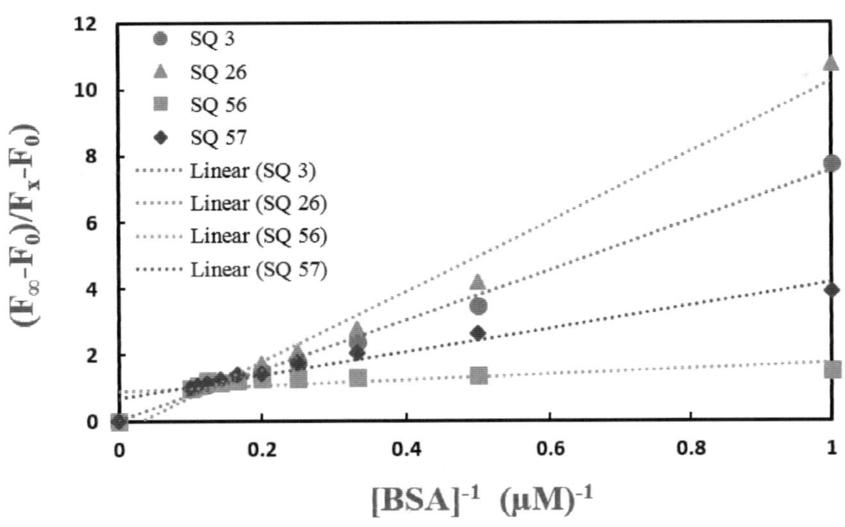

Figure. 6 Plot of ($F_\infty - F_0$) / ($F_X - F_0$) as function $[BSA]^{-1}$ at a fixed dye concentration of 2 μM.

4. Conclusions

Carboxy functionalized symmetrical squaraines with varying substituents were synthesized with satisfactory yields and characterized. The dyes were subjected to photophysical investigations in order to explore their applicability as fluorescent probes. Interactions of these dyes with BSA as a protein model suggested the formation of dye-BSA conjugates, which led to the enhancement in fluorescence intensity along with bathochromic shift of emission maxima. Interaction between dyes with protein was found to be (**SQ-26 > SQ-3 > SQ-57 > SQ-56**) while hydrophobicity of the dyes was found to be in the order (**SQ-57> SQ-3 > SQ-26 > SQ-56**) as confirmed by HPLC retention time. The dye SQ-26 showed highest binding ability with BSA hence showed high binding constant (K_a) which can be attributed to its possible interactions with both the binding sites of BSA owing to its moderate hydrophilicity as well as hydrophobicity. The above photophysical characterization reveals that Squaraine dyes can be used as good fluorophores for designing NIR-FRET systems.

References

[1] Weissleder R and Pittet M J 2008 *Nature.* **452** 580-589.
[2] Kobayashi H, Ogawa M, Alford R, Choyke P L and UranoY 2009 *Chem. Rev.* **110** 2620-2640.
[3] Law K Y 1992 *Chem. Mater.* **4** 605-611.
[4] Chen H, Farahat M S, Law K Y and Whitten D G 1996 *J. Am. Chem. Soc.* **118** 2584-2594.
[5] Ahn H Y, Yao S, Wang X and Belfield K D 2012 *ACS Appl. Mater. Interfaces.* **4** 2847-2854.
[6] McEwen J J and Wallace K J 2009 *Chem. Commun.* 6339-6351.
[7] Ros-Lis J V, García B, Jiménez D, Martínez-Máñez R, Sancenón F, Soto J, Gonzalvo F and Valldecabres M C 2004 *J. Am. Chem. Soc.* **126** 4064-4065.
[8] Kolev T M, Yancheva D Y and Stoyanov S I 2004 Adv. Funct. Mater. **14** 799-805.
[9] Johnson J R, Fu N, Arunkumar E, Leevy W M, Gammon S T, Piwnica-Worms D and Smith B D 2007 Angew. Chem. **119** 5624-5627.
[10] Gassensmith J J, Arunkumar E, Barr L, Baumes J M, DiVittorio K M, Johnson J R, Noll B C and Smith B D 2007 *J. Am. Chem. Soc.* **129** 15054-15059.
[11] Zhang Y Z, Yang Q F, Du H Y, Tang Y L, Xu G Z and Yan W P 2008 *Chin. J. Chem.* **26** 397-401.
[12] W Pham, W F Lai, R Weissleder and C H. Tung 2003 *Bioconjugate chem.* **14** 1048-1051.
[13] Pandey S S, Inoue T, Fujikawa N, Yamaguchi Y and Hayase S 2010 *J. Photochem. Photobiol. A.* **214** 269-275.
[14] Pandey S S, Inoue T, Fujikawa N, Yamaguchi Y and Hayase S 2010 *Thin Solid Films.* **519** 1066-1071.
[15] Zhegalova N G, He S, Zhou H, Kim D M and Berezin M Y 2014 *Contrast Media Mol. Imaging.* **9** 355-362.
[16] Wojtyk J T, Kazmaier P M and Buncel E 2001 *Chem. Mater.* **13** 2547-2551.
[17] Jisha V S, Arun K T, Hariharan M and Ramaiah D 2006 *J. Am. Chem. Soc.* **128** 6024-6025.
[18] Patonay G, Salon J, Sowell J and Strekowski L 2004 Molecules. **9** 40-49.
[19] Welder F, Paul B, Nakazumi H, Yagi S and Colyer C L 2003 *J. Chromatogr. B.* **793**, 93-105.
[20] Suryawanshi V D, Walekar L S, Gore A H, Anbhule P V and Kolekar G B 2016 *J. Pharm. Anal.* **6** 56-63.
[21] Jisha V S, Arun K T, Hariharan M and Ramaiah D 2010 J. Phys. Chem. B. **114** 5912-5919.
[22] Redy O, Kisin-Finfer E, Sella E and Shabat D 2012 Org. Biomol. Chem. **10** 710-715.
[23] Packard, Beverly Z., et al.1997 methods in enzymology. **278** 15-23.
[24] Oushiki D, Kojima H, Takahashi Y, Komatsu T, Terai T, Hanaoka K, Nishikawa M, Takakura Y and Nagano T 2012 *Anal. Chem.* **84** 4404-4410.
[25] Nizomov N, Ismailov Z F, Nizamov S N, Salakhitdinova M K, Tatarets A L, Patsenker L D and Khodjayev G 2006 *J. Mol. Struct.* **788** 36-42.
[26] Jiménez M C, Miranda M A and Vayá I 2005 *J. Am. Chem. Soc.* **127** 10134-10135.

IOP Conf. Series: Journal of Physics: Conf. Series **924** (2017) 012007 doi:10.1088/1742-6596/924/1/012007

Effect of High Pressure on Resistance in Au(dmit)$_2$ LB Films

Y F Miura[1,*,†], **H Hasegawa**[2], **K Torizuka**[3] **and Y Uwatoko**[3]

[1] Toin University of Yokohama
1614 Kurogane-cho, Aoba-ku, Yokohama, 225-8503 JAPAN

[2] KARC, National Institute of Information and Communications Technology (NICT)
588-2 Iwaoka, Nishi-ku, Kobe, 651-2492 JAPAN

[3] Institute for Solid State Physics, University of Tokyo
5-1-5 Kashiwanoha, Kashiwa, 277-8581 JAPAN

*E-mail: yfmiura@hama-med.ac.jp
†Present address: Hamamatsu University School of Medicine, 1-20-1 Handayama, Higashi-ku, Hamamatsu, 431-3192 JAPAN

Abstract. The high pressure effect on resistance of the LB film of ditetradecyldimethylammonium-Au(dmit)$_2$ [$2C_{14}N^+Me_2$-Au(dmit)$_2$] salt has been investigated; the measurements were performed down to 0.8 K and pressures up to 1.03 GPa using a clamp-type pressure cell. The resistance gradually decreases with increasing pressure reaching a 57% smaller value at 1.03 GPa compared to the ambient one. The resistance increases with decreasing temperature for three different conditions of the measurement: at 0 GPa and under pressure (clamped at 0.80 GPa and 1.03 GPa) in the temperature range of 18–295 K. Over the whole temperature range, the higher the pressure is, the lower the sheet resistance is, indicating that the application of pressure is effective to reduce the sheet resistance. Although we have already reported the magnetic anomaly suggesting the Meissner state in the LB film below 3.9 K, no resistance drop is observed even if the resistance was measured down to 0.8 K.

1. Introduction

The Langmuir-Blodgett (LB) technique is a powerful technique that allows us to assemble organic molecules into organized two-dimensional (2D) molecular sheets [1]. Among the various objectives of fabricating functional LB films, the development of highly conductive LB systems is important, because it may open up the possibility of a variety of future applications [2]. So far, several LB systems have been reported to have metallic conducting properties; they are based on either Au(dmit)$_2$ (dmit = 1,3-dithiol-2-thione-4,5-dithiolate) salts [3-7] or bis(ethylenedioxy)tetrathiafulvalene (BEDO-TTF) molecules [8-10].

We have already reported that the LB film based on the ditetradecyldimethylammonium-Au(dmit)$_2$ salt [Fig. 1, $2C_{14}N^+Me_2$-Au(dmit)$_2$ salt] exhibits a high room-temperature conductivity of 40 S/cm with a metallic temperature dependence of resistance after

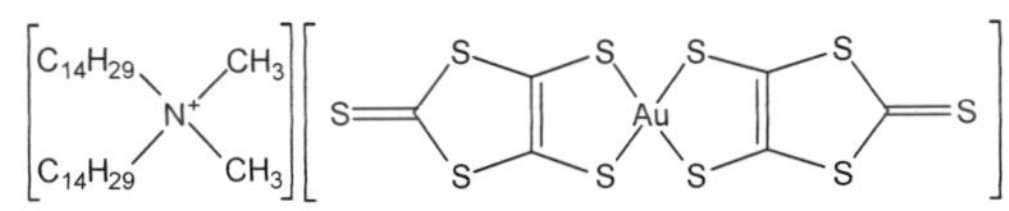

Fig. 1. $2C_{14}N^+Me_2$-Au(dmit)$_2$ salt

electro-oxidation [6,7]. Furthermore, our ac magnetic susceptibility measurement suggests the existence of a superconducting phase below 3.9 K [11]. However, resistance drop on cooling has not yet been observed reproducibly for the LB film.

We have already reported that the $2C_{14}N^{+}Me_2$-Au(dmit)$_2$ LB film consists of domains of 5.20± 1.52 μm [12]. In such an inhomogeneous system, it is possible that the superconductivity is localized and global superconductivity, which shows macroscopic zero resistance, is not realized.

There exist several theoretically-derived threshold resistances for realizing global superconductivity in disordered superconductors. In a disordered 2D system, where superconductivity is competing with Anderson localization, the sheet resistance ($R_{\square}$) should satisfy $R_{\square} < h/e^2 = 25.5$ kΩ, where h = Planck constant and e = elementary charge [13]. In a 2D system, where 2D superconducting domains are linked with thin insulating regions, $R_{\square}$ should satisfy $R_{\square} < h/4e^2 = 6.4$ kΩ, where h = Planck constant and e = elementary charge [14].

The $R_{\square}$value can be reduced by improving the order in the film. But the application of pressure may be another way to reduce the $R_{\square}$value. While the application of high pressure is an established method to induce metallic and superconducting states in various molecular crystals [15], studies on resistance of LB films under pressure are not common and limited to those by ourselves [5, 16, 17]. In this paper, we report on resistance of the LB film of the $2C_{14}N^{+}Me_2$-Au(dmit)$_2$ salt measured under high pressures up to 1.03 GPa using a clamp-type pressure cell.

2. Experimental

2.1. Sample preparation

The ditetradecyldimethylammonium-Au(dmit)$_2$ salt [$2C_{14}N^{+}Me_2$-Au(dmit)$_2$ salt, Fig.1] was synthesized in accordance with the procedure of Steimecke *et al.* [12,18] The $2C_{14}N^{+}Me_2$-Au(dmit)$_2$ salt was dissolved in a 1:1 mixture of benzene and acetonitrile (0.25 mM) and spread on the surface of pure water. After keeping the salts on the water surface for 5 min, they were compressed to 25 mN/m, and then, the floating film was transferred onto a PET (polyethylene terephthalate) substrate (Mitsubishi Plastics, Type S100-100) by the horizontal lifting method (Langmuir-Schaefer method). Twenty layers were deposited onto the substrate. A commercially available LB trough (NIMA Technology Ltd., Type 622) was used.

Prior to the LB film deposition of the $2C_{14}N^{+}Me_2$-Au(dmit)$_2$ salt, the PET substrate, which was 0.1 × 13 × 38 mm^3 in size, was pre-coated with 20-

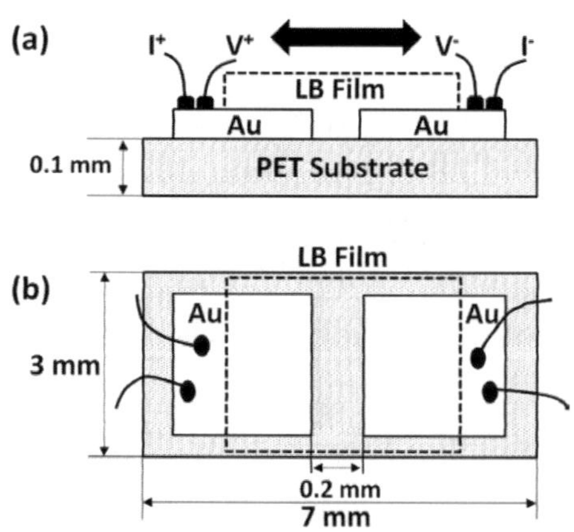

Fig. 2. The schematic representation of the LB film of the $2C_{14}N^{+}Me_2$-Au(dmit)$_2$ salt for resistance measurement under pressure. (a) Cross section; (b) Top view.

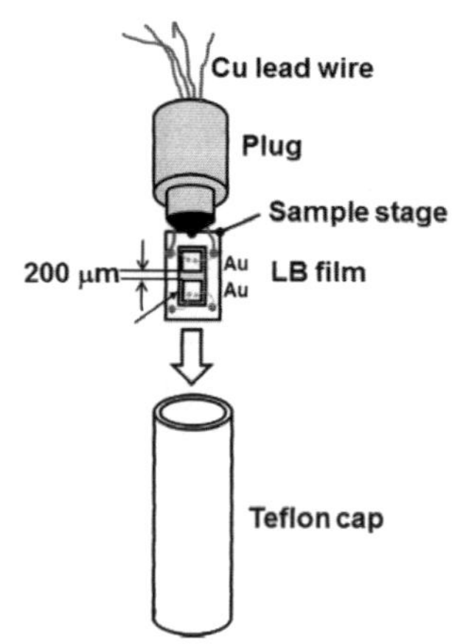

Fig. 3. Experimental setup for applying pressure to the $2C_{14}N^{+}Me_2$-Au(dmit)$_2$ LB film. The LB film attached to the plug is set to the Teflon cap of outer diameter of 8 mm (inner diameter: 7 mm) filled with the pressure medium of Daphne Oil 7373.

layered LB film of cadmium-arachidate, and then, four gold electrode strips with gaps of 0.2 mm were deposited on the substrate by vacuum evaporation. The as-deposited LB film was immersed in $LiClO_4$ (aq.) and electro-oxidized with a constant current of 0.8 μA. The gold electrode underneath the LB layer and a platinum rod set in the electrolyte were used as the working and counter electrodes, respectively. A positive potential was applied to the gold electrode. The details of the sample preparation are in our previous papers [7, 12].

2.2. Resistance measurement under pressure

The electro-oxidized LB film, whose original size was $0.1 \times 13 \times 38$ mm^3, was cut by scissors into a small fraction of $0.1 \times 3 \times 7$ mm^3, as schematically shown in Figs. 2(a) and 2(b). Thin gold wires (50 μm in diameter) were attached to gold electrodes deposited underneath the LB film by a silver paste (Fujikura Kasei Co. Ltd., DOTITE: Type D-550) and the resistance was measured along the film plane by DC or AC two probe method, as shown in Figs. 2 (a) and 2(b).

The LB film with the dimension of $0.1 \times 3 \times 7$ mm^3 was placed on a sample stage, which is made of a filter paper (Toyo Roshi Kaisha, Ltd., Advantec Type No. 1) vacuum impregnated with STYCAST 1266 epoxy encapsulant (Henkel AG & Co. KGaA), as shown in Fig. 3. The sample stage was further attached to a plug, which was made from beryllium copper, using STYCAST 2850 FT epoxy encapsulant (Henkel AG & Co. KGaA), and then, copper wires (0.1 mm in diameter) were attached to the gold wires using the silver paste, as shown in Fig. 3. The plug with the LB film was attached to a Teflon cap filled with Daphne Oil 7373 (Idemitsu Kosan Co. Ltd.).

Fig. 4. Photo of the cylinder body (B) with the screw attached for clamp (A), which have been set at the stage of hydraulic press.

The sample holder, in which the LB sample was set using the pressure medium of Daphne Oil 7373, was inserted in a beryllium copper cylinder body (indicated as "B" in Fig. 4) with a lock nut for clamp (indicated as "A" in Fig.4), and placed on the stage of hydraulic press, as shown in Fig. 4. The pressure was gradually applied to the clamp-type pressure cell using the hydraulic press and clamped by the lock nut at 298 K and it was brought to cryostat to measure temperature dependence of resistance. The temperature dependence of resistance of the LB film cut from the same sample batch was also measured at 0 GPa (in vacuum) for comparison.

3. Results and Discussion

The resistance was first measured with increasing the pressure from ambient pressure up to 0.80 GPa at 298 K. Then, the pressure cell was clamped at 0.80 GPa and was cooled down to measure the temperature dependence of resistance. After the pressure cell was warmed up to 298 K, the resistance was further measured with increasing the pressure up to 1.03 GPa. Finally, the pressure cell was clamped at 1.03 GPa and was cooled down to measure the temperature dependence of resistance again [19].

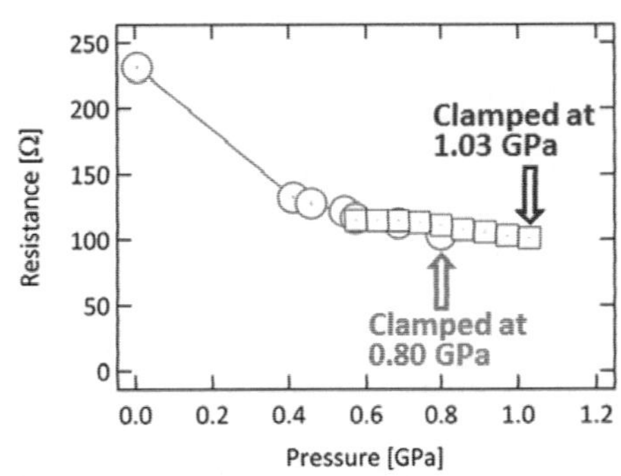

Fig. 5. Resistance at 298 K plotted against pressure; the first and secondary sequences of measurements are shown by open circles and squares, respectively.

Figure 5 shows the resistance of the $2C_{14}N^{+}Me_2$-$Au(dmit)_2$ LB film plotted against pressure; the

IOP Conf. Series: Journal of Physics: Conf. Series **924** (2017) 012007 doi:10.1088/1742-6596/924/1/012007

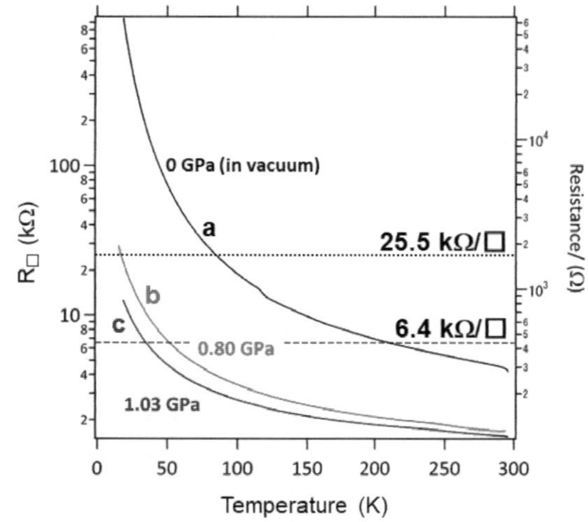

Fig. 6(a). Sheet resistance (R□), plotted against temperature. (a) Measured at 0 GPa (in vacuum); (b) 0.80 GPa (clamped at 298 K); (c) 1.03 GPa (clamped at 298 K).

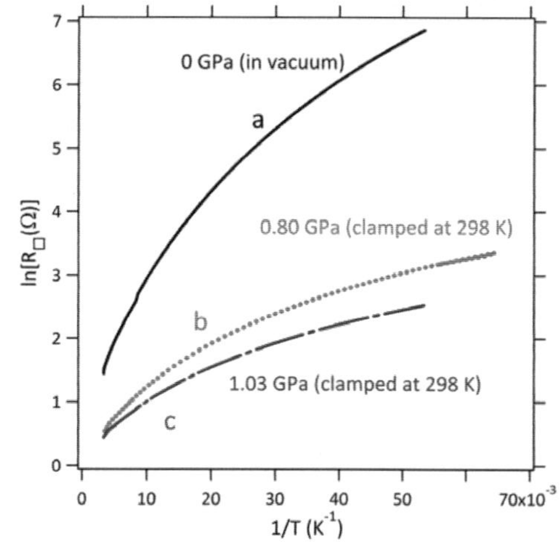

Fig. 6(b). The Arrhenius plot of the sheet resistance (R□). (a) Measured at 0 GPa (in vacuum); (b) 0.80 GPa (clamped at 298 K); (c) 1.03 GPa (clamped at 298 K).

resistance was measured along the film plane by DC two probe method at 298 K. The resistance gradually decreases with increasing pressure reaching a 57% smaller value at 1.03 GPa compared to the ambient one.

Figure 6(a) shows the temperature dependence of sheet resistance of the $2C_{14}N^{+}Me_2$-Au(dmit)$_2$ LB film measured at 0 GPa and under pressure (clamped at 0.80 GPa and 1.03 GPa) by AC two probe method (17 Hz). Resistance increases with decreasing temperature for all of the three different conditions of the measurement in the temperature range of 18−295 K, as shown in Fig. 6(a). Furthermore, over the whole temperature range, the higher the pressure is, the lower the sheet resistance is, indicating that the application of high pressure to the LB film is effective to reduce the sheet resistance.

We have already reported that the $2C_{14}N^{+}Me_2$-Au(dmit)$_2$ LB film shows metallic temperature dependence down to a certain temperature when measured at 0 GPa [6,7]. However, it should be noted here that some sample batches do not show the metallic behaviour when the electrode gap for the measurement is as wide as 0.2−0.5 mm; this is possibly due to disorders such as grain boundaries and defects. As for this sample batch, the metallic behaviour is not observed at 0 GPa and the pressure application does not induce the metallic behaviour, although it reduces the sheet resistance, as shown in Figs 6(a).

Figure 6(b) shows the Arrhenius plot of sheet resistance (R□). Logarithm of R□ plotted against the inverse of temperature is sublinear for all of the three different conditions of the measurement: measured at 0 GPa and under pressure (clamped at 0.80 GPa and 1.03 GPa). Since the pressure in the clamp-type cell gradually decreases on cooling, the interpretation of the temperature dependence of resistance is not straightforward. For instance, the pressure at around 7 K is *ca.* 20% lower than that at 298 K, when Daphne Oil 7373 was used as the pressure medium [20]. However, we hypothesise that the resistance behaviours under pressure, which are deviated from the simple activation-type one, are intrinsic because the decrease of pressure on cooling would give a superlinear behaviour for the sample with the simple activation-type property and resistance measured in vacuum also has a similar sublinear behaviour.

Deviations from the simple activation-type behaviour have been reported for several conductive LB films. In those literatures, at lower temperature regions, 2D variable range hopping (VRH) regime [4], thermal-fluctuation-induced tunnelling (TFIT)[5], and 2D weak localization [20] have been taken into account to interpret the resistance behaviours in the heterogeneous LB systems. As for the present LB system, more investigations are necessary to elucidate the conduction mechanism.

As stated in introduction section, for disordered superconductors, there exit some theoretical values of threshold sheet resistance ($R_\square$) for realizing global superconductivity. In Fig. 6(a), the two different thresholds are shown by dotted line ($R_\square < h/e^2 = 25.5$ kΩ) and dashed line ($R_\square < h/4e^2 = 6.4$ kΩ), which are the cases when superconductivity is competing with Anderson localization [13] and when 2D superconducting domains are linked with thin insulating regions [14], respectively.

We have tentatively performed the resistance measurement of the $2C_{14}N^{+}Me_2$-Au(dmit)$_2$ LB film in the temperature range of 0.8−18 K at ambient pressure and under pressure (clamped at 0.80 GPa and 1.03 GPa), although the data are not shown in Fig. 6(a). None of the $R_\square$values of the three different conditions of measurement satisfies those two thresholds at the lower temperatures and no resistance drop is seen even if the measurement has been performed down to 0.8 K. We consider that it is necessary to reduce the $R_\square$values by improving the order in the LB film and applying higher pressures.

4. Conclusion and Future Directions

The resistance of the $2C_{14}N^{+}Me_2$-Au(dmit)$_2$ LB film decreases with increasing pressure at room temperature; it reaches a 57% smaller value at 1.03 GPa compared to the ambient one. The temperature dependence of resistance of the LB film was measured at ambient pressure and under pressure (clamped at 0.80 GPa and 1.03 GPa). In the temperature range of 18−295 K, the higher the pressure is, the lower the sheet resistance is, indicating that the application of high pressure to the LB film is an effective method to reduce the sheet resistance ($R_\square$). All the three behaviours of the temperature dependence of the sheet resistance are of sublinear. Since the decrease of pressure inside the pressure cell on cooling cannot explain the sublinear behaviour in the Arrhenius plot, we hypothesise that it reflects an intrinsic property of conduction in the film and studies on electronic transport properties in the LB film are now in progress.

As stated in introduction section, we have already reported that the ac magnetic susceptibility of the $2C_{14}N^{+}Me_2$-Au(dmit)$_2$ LB film suggests the existence of the Meissner state below 3.9 K. At this stage, however, the origin of the magnetic anomaly of the LB film is still an open question. One of the hypotheses to interpret the data is that superconductivity exists locally in the heterogeneous LB system, but due to disorders, global superconductivity is not observed. Studies are now in progress to reduce the $R_\square$values by improving the order in the LB film and applying higher pressures toward the realization of global superconductivity in the LB film.

Acknowledgment

This work was supported in part by JSPS KAKENHI Grant No. 15K04655. We would like to thank Prof. Masato Hedo and Prof. Kazuyuki Matsubayashi for helpful guidance in resistance measurement under high pressure in early stages of this work.

References

[1] Ulman A 1991 *An Introduction to Ultrathin Organic Films: from Langmuir-Blodgett to Self-Assembly* (San Diego, CA: Academic Press)
[2] Nakamura T 1997 *Handbook of Organic Conductive Molecules and Polymer*, ed. Nalwa H S, vol 1, Chap. 14 (New York: Wiley)

IOP Conf. Series: Journal of Physics: Conf. Series **924** (2017) 012007 doi:10.1088/1742-6596/924/1/012007

[3] Nakamura T, Kojima K, Matsumoto M, Tachibana H, Tanaka M and Kawabata Y 1989 *Chem Lett* **18** 367.
[4] Miura Y F, Takenaga M, Kasai A, Nakamura T, Matsumoto M and Kawabata Y 1991 *Jpn J Appl Phys* **30** 3503
[5] Isotalo H, Paloheimo J, Miura Y F, Azumi R, Matsumoto M and Nakamura T 1995 *Phys Rev B* **51** 1809.
[6] Miura Y F, Okuma Y, Ohnishi H and Sugi M 1998 *Jpn J Appl Phys* **37** L1481
[7] Miura Y F, Kitao M, Matsui H, Sugi M, Hedo M, Matsubayashi K and Uwatoko Y 2008 *Jpn J Appl Phys* **47** 8884
[8] Nakamura T, Yunome G, Azumi R, Tanaka M, Matsumoto M, Horiuchi S, Yamochi H and Saito G 1994 *J Phys Chem* **98** 1882
[9] Ogasawara K, Ishiguro T, Horiuchi S, Yamochi H and Saito G 1996 *Jpn J Appl Phys* **35** L571
[10] Ohnuki H, Noda T, Izumi M, Imabukuro T and Kato R 1997 *Phys Rev B* **55** R10225
[11] Miura Y F, Horikiri M, Saito S-H and Sugi M 2000 *Solid State Commun* **113** 603
[12] Miura Y F, Matsui H, Inoue K, Hoshino J and Ikegami K 2015 *Synthetic Metals* **207** 54
[13] Maekawa S and Fukuyama H 1981 *J Phys Soc Jpn* **51** 1380
[14] Fisher M P A 1990 *Phys Rev B* **36** 1917
[15] Ishiguro T, Yamaji K and Saito G (Eds.) 1997 *Organic Superconductors, 2nd ed.* (Berlin: Springer)
[16] Miura Y F, Isotalo H, Matsumoto M and Nakamura T 1993 *Appl Phys Lett* **63** 1705
[17] Miura Y F, Morita S, Saito S-H, Sugi M, Hedo M and Uwatoko Y 2002 *Int'l J Nanoscience* **1** 627
[18] Steimecke G, Sieler H J, Kirmse P and Hoyer E 1979 *Phosphorus Sulfur* **7** 49
[19] We have confirmed that the resistance behaviors measured by the dc or ac two-probe method are in good agreement with those measured by the dc or ac four-probe method. The effect of contact resistance is considered to be negligible when the resistance of the ultra-thin film is more than 10^2-10^3 Ω as in the case shown in Figures 6(a) because the contact resistance is assumed to be around 1 Ω. This has been confirmed for the LB film under which four electrode strips are formed (please see Figure S1 in supplementary data), although it has been only confirmed for the ambient measurement. Measurements of the resistance under pressure using the dc or ac four-probe method has not yet been performed due to the limitation of the sample size but it is now in progress and will be published in a separate paper.
[20] Pressure at lower temperature in the clamp-type cell has been estimated utilizing the linear pressure dependence of critical temperature of superconductivity of lead [21]. The pressure at around 7 K is estimated as 0.64 GPa when it is clamped at 0.80 GPa at 298 K and it is estimated as 0.84 GPa when it is clamped at 1.03 GPa at 298 K.
[21] Boughton, R I, Olsen J L and Palmy C 1970 *Progress in Low Temperature Physics*, ed. Gorter C J, Vol 6, Chap. 4 (Amsterdam: North-Holland)

Solution-processed organic field-effect transistors based on dinaphthothienothiophene precursor with chemically modified electrodes

Takashi Nagase[1,2], Souichiro Abe[1], Takashi Kobayashi[1,2], Yu Kimura[3], Azusa Hamaguchi[3], Yoshinori Ikeda[3] and Hiroyoshi Naito[1,2]

[1] Department of Physics and Electronics, Osaka Prefecture University, Sakai 599-8531, Japan
[2] The Research Institute for Molecular Electronic Devices (RIMED), Osaka Prefecture University, Sakai 599-8531, Japan
[3] Electronics Materials Development Project, Teijin Limited, Hino, Tokyo 191-8512, Japan

E-mail: nagase@pe.osakafu-u.ac.jp, yos.ikeda@teijin.co.jp, naito@pe.osakafu-u.ac.jp

Abstract. Bottom-gate organic field-effect transistors (OFETs) based on a soluble precursor of dinaphtho[2,3-b:2',3'-f]thieno[3,2-b]thiophene (DNTT) with bottom-contact source-drain electrodes chemically modified with pentafluorobenzenethiol (PFBT) have been fabricated. The preparation of DNTT films using CYTOP overcoat layers allows the solution processing of bottom-gate/bottom-contact DNTT FETs with good electrical contacts between the PFBT-treated Au electrodes and the DNTT molecules. The DNTT FETs processed using CYTOP overcoat layers exhibit the field-effect mobilities of up to 0.37 $cm^2\ V^{-1}\ s^{-1}$. High maximum mobility of 0.29 $cm^2\ V^{-1}\ s^{-1}$ has been achieved in solution-processed DNTT FETs with channel length of 5 μm.

1. Introduction

In recent years, a highly π-extended small-molecule semiconductor, dinaphtho[2,3-b:2',3'-f]thieno[3,2-b]thiophene (DNTT) [1], has attracted considerable interest as a promising channel material for high-performance organic field-effect transistors (OFETs). OFETs with vacuum-deposited DNTT thin films exhibit high field-effect mobilities of up to 2.9 $cm^2\ V^{-1}\ s^{-1}$ and good air stability because the DNTT enables thin-film formation with excellent microstructures and has a high ionization potential (5.4 eV) [1 - 3]. Recent interest has also been directed towards the development of high-mobility solution-processable OFETs. However, DNTT molecules show low solubility in organic solvents, even when long alkyl chains are attached to the molecules [4]. A DNTT precursor, 5,14-*N*-phenylmaleimide dinaphtho[2,3-b:2',3'-f]thieno[3,2-b]thiophene (Fig. 1), can be dissolved in organic solvents (*e.g.*, chloroform, dichlorobenzene, or dimethylacetamide) at room temperature because of its low crystallinity, and can be converted into DNTT by thermal annealing through the elimination of *N*-phenylmaleimide [5]. We have recently demonstrated the solution processing of high-performance bottom-gate/top-contact DNTT FETs through vertical phase separation that appeared in solution-processed blend films of the DNTT precursor and an insulating polymer [5 - 7].

IOP Conf. Series: Journal of Physics: Conf. Series **924** (2017) 012008 doi:10.1088/1742-6596/924/1/012008

For application of OFETs to organic integrated circuits, the development of bottom-gate/bottom-contact DNTT FETs is desirable since fine-pattering techniques, such as photolithography and ink-jet printing, can be applied to the fabrication of source-drain electrodes with narrow spacing. It is widely known that the field-effect mobility of bottom-gate/bottom-contact OFETs depends sensitively on a charge injection barrier formed at the source electrode/organic semiconductor interface as compared with that of bottom-gate/top-contact OFETs [8 - 10]. The chemical modification of Au source-drain electrodes with a thiolated self-assembled monolayer (SAM) of pentafluorobenzenethiol (PFBT) has been found to increase the work function of Au electrodes (~4.8 eV) to 5.2-5.5 eV [11, 12]. This leads to the enhancement in the field-effect mobility of bottom-gate/bottom-contact short-channel OFETs with vacuum-deposited alkylated DNTT thin films as a result of the decrease in the hole injection barrier (contact resistance) at the Au electrode/DNTT interface [13]. However, we observed that the solution processing of DNTT films on substrates with PFBT-treated Au electrodes does not allow forming continuum films at boundaries between the substrate and the PFBT-treated electrodes.

In this study, we develop a simple technique for fabricating bottom-gate/bottom-contact OFETs based on the DNTT precursor with PFBT-treated source-drain electrodes by using an overcoat layer of CYTOP. The DNTT FETs processed using CYTOP overcoat layers exhibit higher electrical performances as compared with those of devices processed without using CYTOP overcoat layers. Solution-processed DNTT FETs exhibit high field-effect mobilities of up to 0.37 $cm^2 V^{-1} s^{-1}$.

2. Experimental

Figure 1 shows the schematic diagram of a bottom-gate/bottom-contact DNTT FET device processed using the DNTT precursor. Au source-drain electrodes (40 nm) with thin Cr adhesion layers (2 nm) were fabricated by vacuum evaporation through a metal shadow mask on a heavily doped Si wafer with a SiO_2 layer (300 nm), which were used as the gate electrode and the gate insulator, respectively. The substrates with Cr/Au electrodes were washed sequentially with acetone and 2-propanol in ultrasonic baths and were then cleaned with O_2 plasma to remove organic contaminants. The surfaces of Au electrodes were chemically modified with the SAM of PFBT (Tokyo Chemical Industry Co., Ltd.) to reduce a hole injection barrier caused by the potential difference between the work function of Au (~4.8 eV) and the ionization potential of DNTT (5.4 eV). To form the SAMs on Au surfaces, the substrates were immersed in a ~50 mM ethanol solution of PFBT overnight at room temperature in a capped bottle. After removing the substrates from the PFBT solution, they were ultrasonically washed in fresh ethanol to remove any excess layers. A 0.2 wt% chloroform solution of the DNTT precursor, which was synthesized in our laboratory [5], was drop cast on the substrates and dried in ambient air. Then, an amorphous floropolymer solution of CYTOP (CTL-809M, Asahi Glass Co., Ltd) was spin-coated on the DNTT precursor films and subsequently annealed at 200 °C for 5 min in air to convert DNTT precursor into DNTT. We also fabricated OFETs with DNTT films on substrates having PFBT-treated Cr/Au electrodes (2 nm/40 nm) without using CYTOP overcoat layers for comparison. The channel length (L) and width (W) of these FET devices were 350 μm and 2 mm, respectively.

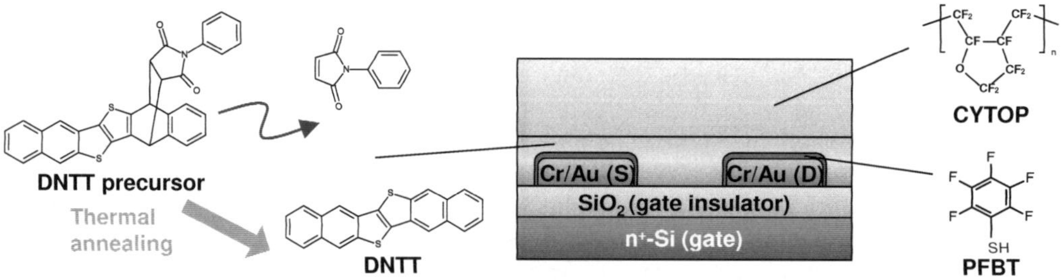

Figure 1. Chemical structure of organic materials and the schematic diagram of a bottom-gate/bottom-contact OFET processed using the DNTT precursor and the CYTOP overcoat layer.

To investigate the dependence of field-effect mobilities on channel length, we fabricated bottom-gate/bottom-contact DNTT FETs with *L*s ranging from 250 to 5 μm and a fixed *W* of 1.5 mm using photolithographically-defined Cr/Au source-drain electrodes on Si/SiO_2 substrates, where the surfaces of SiO_2 gate insulators were treated with a SAM of phenethyltrichlorosilane (PETS) (Gelest, Inc.) to suppress hysteresis behavior for the forward and reverse scans of the gate voltage often reported for *p*-type OFETs with SiO_2 insulators [14, 15]. The substrates were immersed in a ~50 mM toluene solution of PETS overnight after PFBT treatment of Au electrodes to form the PETS SAMs on SiO_2 surfaces.

The current-voltage characteristics of OFET devices were measured in a nitrogen-filled glove box or vacuum using source meters (Keithley 6430, 2635, and 2400) at room temperature.

3. Results and discussion

Figures 2(a) and 2(b) show the polarized optical micrographs of regions around the edges of PFBT-treated Au electrodes in DNTT FETs fabricated without and with CYTOP overcoat layer, respectively. Solution-processing DNTT FETs with PFBT-treated Au electrodes without using CYTOP layers causes needle-like crystals around the electrode edges, whereas firm forming with CYTOP layers suppresses the formation of needle-like crystals. The typical gate voltage (V_G)-drain current (I_D) (transfer) and drain voltage (V_D)-I_D (output) characteristics of both DNTT FETs are shown in Figs. 2(c) and 2(d), respectively. It is found that the drain current of devices is remarkably increased using CYTOP overcoat layers, indicating better connectivity between DNTT molecules and PFBT-treated electrodes in devices with CYTOP overcoat layers.

The field-effect mobility (μ) was determined in the saturation regime from the slope of $|I_D|^{1/2}$-V_G curve in the low V_G region using the standard FET equation: $I_D = (W/2L)\mu C_i(V_G - V_{th})^2$, where C_i is the capacitance per unit area for the gate insulator and V_{th} is the threshold voltage. The maximum field-effect mobility of PFBT-treated devices with CYTOP layers was 0.37 $cm^2\ V^{-1}\ s^{-1}$, which is much higher than that of PFBT-treated devices without CYTOP layers (0.03 $cm^2\ V^{-1}\ s^{-1}$) and that of non-treated devices with CYTOP layers (0.04 $cm^2\ V^{-1}\ s^{-1}$). The average mobility was 0.17 $cm^2\ V^{-1}\ s^{-1}$, which is comparable to or higher than the mobilities of bottom-gate/top-contact devices fabricated on non-treated SiO_2 gate insulators (0.085-0.13 $cm^2\ V^{-1}\ s^{-1}$ [5]). These results clearly indicate that the PFBT treatment effectively works to reduces hole injection barriers at the Au source electrode/DNTT semiconductor interfaces.

The exact reason for the formation of needle-like crystals for DNTT FETs without CYTOP layers

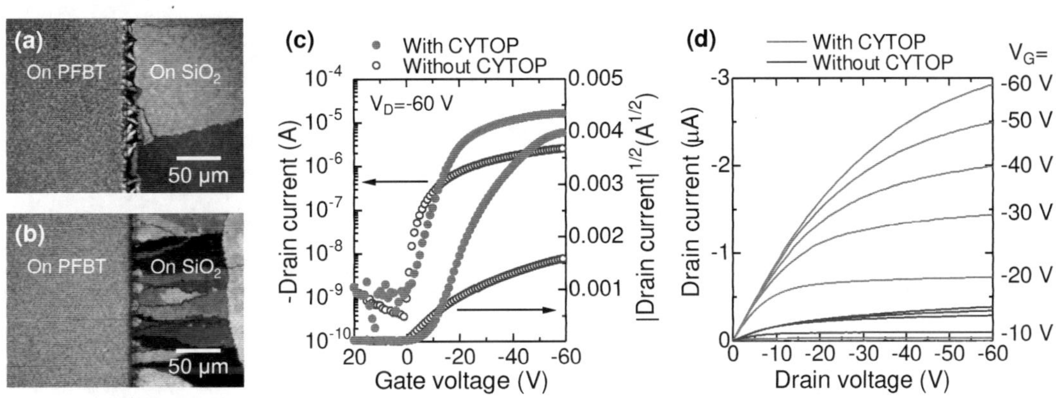

Figure 2. Polarized optical micrographs of regions around the PFBT-treated Au electrodes in bottom-gate/bottom-contact OFETs based on the DNTT precursor fabricated (a) without and (b) with CYTOP overcoat layers. (c) Transfer and (d) output characteristics of both DNTT FETs without and with CYTOP overcoat layers. The channel length of devices was 350 μm.

is unclear at present. However, the enhancement in the growth of highly crystalline films on low energy surfaces reported for solution-processed OFETs [16] could be responsible for the formation of needle-like crystals because PFBT-treated Au electrodes have high hydrophobicity (its water contact angles was ~85°) [17]. In fact, it has been found that PFBT-treated Au electrodes induce the crystallization of a soluble small-molecule semiconductor and the growth of crystalline films extends into the channel region [18]. We consider that the confinement of DNTT molecules between the substrate and the CYTOP overcoat layer would facilitate two-dimensional crystal growth of DNTT films during thermal annealing.

Figures 3(a) and 3(b) show the representative transfer and output characteristics of solution-processed DNTT FETs using CYTOP overcoat layers with different channel lengths, respectively. In these devices, PFBT and PETS were employed for SAM treatment of the Au electrodes and the SiO_2 insulator, respectively. The devices exhibit good transistor performances with high on/off ratios of over 10^5 and small hysteresis in the transfer characteristics between the forward and backward scans of the gate voltage. The hysteresis behavior of *p*-type OFETs with SiO_2 gate insulators is believed to be caused by long-lifetime trapping of electrons by hydroxyl (OH) groups existed at the SiO_2 surfaces [14,

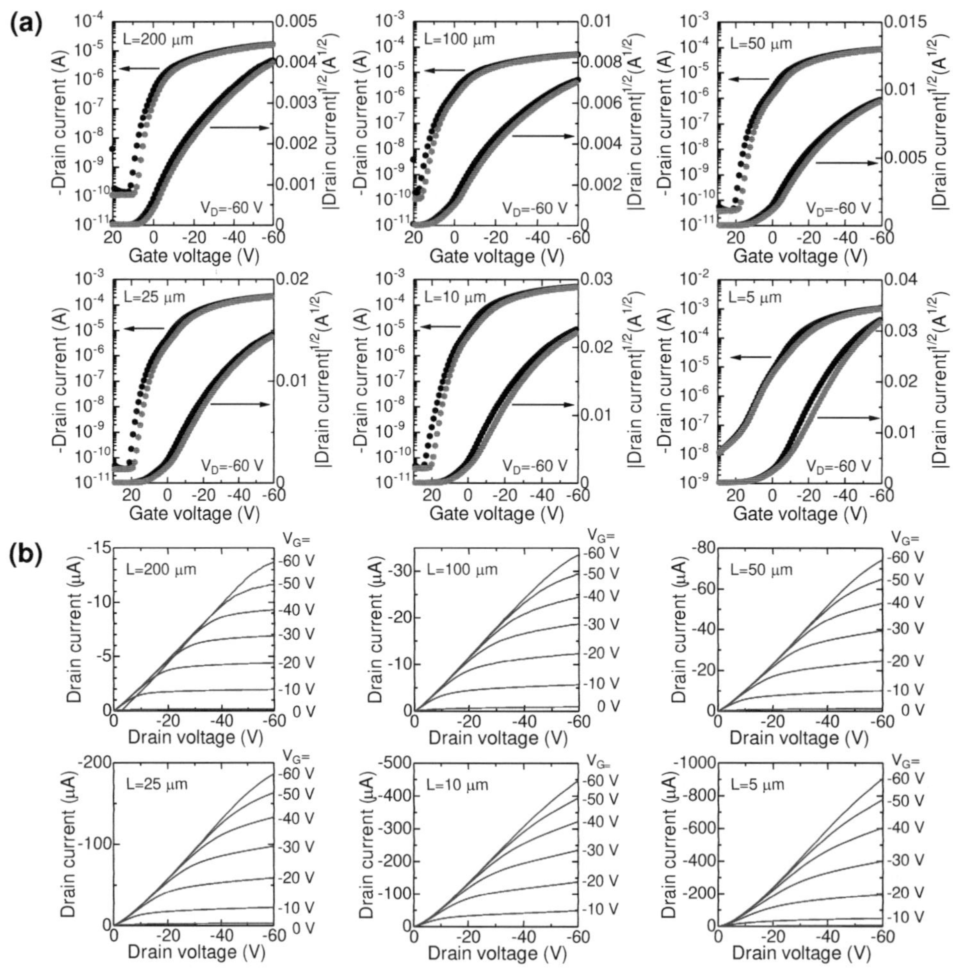

Figure 3. (a) Transfer and (b) output characteristics of solution-processed bottom-gate/bottom-contact DNTT FETs with channel lengths of 200 μm, 100 μm, 50 μm, 25 μm, 10 μm, and 5 μm.

IOP Conf. Series: Journal of Physics: Conf. Series **924** (2017) 012008 doi:10.1088/1742-6596/924/1/012008

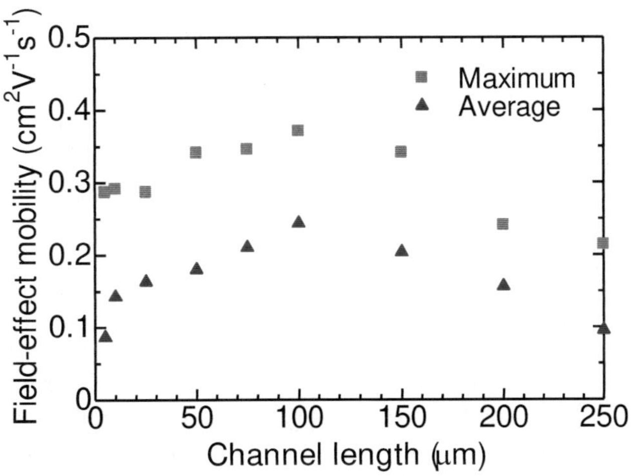

Figure 4. Channel length-dependence of the field-effect mobility of solution-processed bottom-gate/bottom-contact DNTT FETs.

15], and therefore the observed small hysteresis in our devices can be attributed to the termination of the OH groups with PETS SAMs.

The dependence of field-effect mobilities on channel length of devices is shown in Fig. 4. It is found that the field-effect mobility of solution-processed bottom-gate/bottom-contact DNTT FETs tends to increase as the channel length is reduced from 250 to 100 μm. Such behavior can be understood by the decrease in the boundaries of crystalline DNTT domains between source-drain electrodes as shown in the polarized optical micrographs of channel regions (Fig. 5), since it can reduce frequent carrier trapping at the domain boundaries between source-drain electrodes.

We also see from Fig. 4 that the field-effect mobility gradually decreases as the channel length is reduced from 100 to 5 μm. This is likely caused by the influence of contact resistance, and the decrease in the total resistance of devices associated with the reduction of channel length is restricted in case of the presence of high contact resistance, which leads to the apparent decrease in the field-effect mobility in short-channel devices [19]. However, the solution-processed DNTT FETs with channel length of 5 μm achieved a high field-effect mobility of 0.29 $cm^2 V^{-1} s^{-1}$ at maximum. It has been observed that the field-effect mobility of solution-processed DNTT FETs depends on the film annealing temperature and time [5]. The contact resistances of OFETs are also known to be reduced by the increase in the gate voltage. The optimization of film processing processes and the use of thinner gate insulators, such as SAM-treated oxidized Al (AlO_x) insulators [2, 3], can therefore lead to further improvement in the field-effect mobilities of solution-processed DNTT FETs.

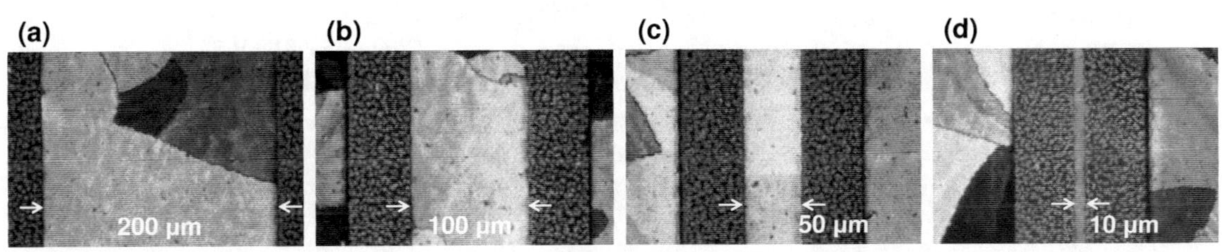

Figure 5. Polarized optical micrographs of the channel region of solution-processed DNTT FETs with channel lengths of (a) 200 μm, (b) 100 μm, (c) 50 μm, and (d) 10 μm.

4. Conclusions

We have examined the solution processing of bottom-gate DNTT FETs based on the DNTT precursor with bottom-contact source-drain electrodes chemically modified with fluorinated PFBT SAMs and investigated their electrical performances. It is found that the use of CYTOP overcoat layers in forming DNTT films allows fabricating bottom-gate/bottom-contact DNTT FETs with high field-effect mobilitites of up to 0.37 $cm^2\,V^{-1}\,s^{-1}$. The field-effect mobility tends to increase with reducing channel length and the devices with channel length of 5 μm exhibit a field-effect mobility of 0.29 $cm^2\,V^{-1}\,s^{-1}$ at maximum.

Acknowledgements

This work is supported by a Grant-in-Aid for Scientific Research on Innovative Areas "New Polymeric Materials Based on Element Blocks" (No. JP24102011) from the Ministry of Education, Culture, Sports, Science, and Technology, Japan. This work is partly supported by a Grant-in-Aid for Scientific Research (C) (No. JP26420276) from the Japan Society for the Promotion of Science, Japan. The authors would like to thank the Nippon Kayaku Corporation for providing the DNTT material.

References

[1] Yamamoto T and Takimiya K 2007 *J. Am. Chem. Soc.* **129** 2224
[2] Zschieschang U, Ante F, Yamamoto T, Takimiya K, Kuwabara H, Ikeda M, Sekitani T, Someya T, Kern K and Klauk H 2010 *Adv. Mater.* **22** 982
[3] Zschieschang U, Ante F, Kälblein D, Yamamoto T, Takimiya K, Kuwabara H, Ikeda M, Sekitani T, Someya, Blochwitz-Nimoth J and Klauk H 2011 *Org. Electron.* **12** 1370
[4] Kang M J, Doi I, Mori H, Miyazaki E, Takimiya K, Ikeda M and Kuwabara H 2011 *Adv. Mater.* **23** 1222
[5] Kimura Y, Nagase T, Kobayashi T, Hamaguchi A, Ikeda Y, Shiro T, Takimiya K and Naito N 2015 *Adv. Mater.* **27** 727
[6] Hamaguchi A, Negishi T, Kimura Y, Ikeda Y, Takimiya K, Bisri S Z, Iwasa Y and Shiro T 2015 *Adv. Mater.* **27** 6606
[7] Kimura Y, Hamaguchi A, Ikeda Y, Nagase T, Naito H, Takimiya K and Shiro T 2015 *Appl. Phys. Express* **8** 101601
[8] Hill I G 2005 *Appl. Phys. Lett.* **87** 163505
[9] Shim G -H, Maruoka F and Hattori R 2010 *IEEE Trans. Electron. Dev.* **57** 195
[10] Marinkovic M, Belaineh D, Wagner V and Knipp D 2012 *Adv. Mater.* **24**, 4005
[11] Kitamura M, Kuzumoto Y, Kang W, Aomori S and Arakawa Y 2010 *Appl. Phys. Lett.* **97** 033306
[12] Kuzumoto Y and Kitamura M 2014 *Appl. Phys. Express* **7** 035701
[13] Kitamura M, Kuzumoto Y and Arakawa Y 2013 *Phys. Status Solidi* C**10** 1632
[14] Gu G, Kane M G, Doty J E and Firester A H 2005 *Appl. Phys. Lett.* **87** 243512
[15] Kim W J, Kim C S, Jo S J, Lee S W, Lee S J and Baik H K 2007 *Electrochem. Solid-State Lett.* **10** H1
[16] Kline R J, McGehee M D and Toney M F 2006 *Nature Mater.* **5** 222
[17] Tatara S, Kuzumoto Y and Kitamura M 2016 *Japan J. Appl. Phys.* **55** 03DD02
[18] Gundlach D J, Royer J E, Park S K, Subramanian S, Jurchescu O D, Hamadani B H, Moad A J, Kline R J, Teague L C, Kirillov O, Richter C A, Kushmerick J G, Richter L J, Parkin S R, Jackson T N and Anthony J E 2008 *Nature Mater.* **7** 216
[19] Klauk H, Schmid G, Radlik W, Weber W, Zhou L, Sheraw C D, Nichols J A and Jackson T N 2003 *Solid-State Electron.* **47** 297

IOP Conf. Series: Journal of Physics: Conf. Series **924** (2017) 012009 doi:10.1088/1742-6596/924/1/012009

Relation between active-layer thickness and power conversion efficiency in P3HT:PCBM inverted organic photovoltaics

S. Nakami[1], T. Narioka[1], T. Kobayashi[1,2], T. Nagase[1,2], H. Naito[1,2]

[1] Department of Physics and Electronics, Osaka Prefecture University, 1-1 Gakuencho, Naka, Sakai, 599-8531 Japan.
[2] Research Institute of Molecular Electronic Devices (RIMED), Osaka Prefecture University, 1-1 Gakuencho, Naka, Sakai, 599-8531 Japan.

E-mail: tkobaya@pe.osakafu-u.ac.jp (TK); naito@pe.osakafu-u.ac.jp (HN)

Abstract. The dependence of active-layer thickness on the power conversion efficiency (PCE) of inverted organic photovoltaics (OPVs) based on poly(3-hexylthiphene) and [6,6]-phenyl-C_{61}-butyric acid methyl ester was investigated. When PCEs were measured immediately after device fabrication, the optimum thickness was ~100 nm. It was, however, found that thick OPVs exhibit higher PCEs a few months later, whereas thin OPVs simply degraded with time. Consequently, the optimum thickness changed with time. Considering this fact, we discuss the relationship between the active-layer thickness and PCE.

1. Introduction

Organic photovoltaics (OPVs) based on conjugated polymers and fullerene derivatives have been extensively studied. As the active-layer thickness increases, the number of absorbed photons increases but the charge collection efficiency is expected to reduce. These two opposite changes determine the relation between active-layer thickness and power conversion efficiency (PCE). This relationship has been experimentally and theoretically investigated by many researchers [1–14]. However, the reported experimental results are highly variable. For example, the optimum thickness for achieving the best PCE has been reported to be in the range of 85–830 nm even when the same materials were used, *e.g.*, poly(3-hexylthiophene) (P3HT) and [6,6]-phenyl-C_{61}-butyric acid methyl ester (PCBM) [1, 2, 7–11]. The reason for this large variation in the optimum thickness may be the high sensitivity of PCE to processing conditions; however, this has not yet been fully clarified so that the true thickness dependence of PCE still remains unclear.

Recently, we found that thick inverted OPVs aged in the dark for a few months exhibit a higher PCE than OPVs that are measured immediately after device fabrication. As opposed to the PCE of thick inverted OPVs, that of thin inverted OPVs simply degrades with time. Consequently, the optimum thickness increases with time. As mentioned below, this change in PCE with time suggests that a processing condition leading to the best PCE for a particular thickness is not necessarily suitable for a different thickness. In short, a processing condition has to be optimized for each thickness. However, in many previous reports, OPVs with various thicknesses were fabricated with identical processing conditions. We thus conjecture that the large variation of the optimum thickness in literature is related to the use of an identical processing condition. In this report, we first present the change in PCE with time in inverted OPVs with various thicknesses and discuss the mechanism of the change. Then, by considering the observed change in PCE, we also discuss the true thickness dependence of PCE.

2. Materials and Methods

P3HT and PCBM were blended in a weight ratio of 1:0.8 and dissolved in chlorobenzene. The concentrations of the solutions were varied in the range of 1.8–5.6 wt% to change the resultant active-layer thickness. The solutions were stirred overnight to promote complete dissolution. ZnO nanoparticle ink was spin-coated onto cleaned ITO-coated substrates, and the substrates were then annealed at 200 °C for 10 min in an ambient atmosphere. Following the annealing process, the substrates were transferred into a glove box filled with dry nitrogen gas, and the active layer was spin-coated from the chlorobenzene solutions onto the ZnO surface at 800 rpm. The films were dried in vacuum for 1 h and subsequently annealed at 150 °C for 10 min unless otherwise noted. As the anode, MoO_3 and Al, instead of Au, was deposited in vacuum at 2×10^{-3} torr. Compared with Au, Al has a higher reflectivity in the visible light region. Finally, the device was sealed by covering with a piece of glass and fixing it with epoxy. The current–voltage characteristics and PCE were measured under a simulated irradiation of 100 mW/cm^2 (AM 1.5 G) from a solar simulator. The samples were stored in the dark in an ambient atmosphere. Most of the aged samples were prepared by storing for 2 months, but some were prepared by storing for 3 months. The difference in the storing time was due to the uncontrollable, uneven degradation time.

3. Results and Discussion

Figure 1 shows the thickness dependence of PCE for inverted OPVs. In the case of fresh inverted OPVs, the optimum thickness is around 100 nm, above which PCE gradually decreases with increasing active-layer thickness. This result is consistent with one report on inverted OPVs [9] and some on conventional OPVs [1, 7]. When measured 2–3 months after device fabrication, PCEs of the inverted OPVs change depending on the thickness. As shown in Fig. 1, PCEs simply degrade at 110 and 160 nm, whereas PCEs improve in the thicker region above 430 nm. In Fig. 2, we show the current–voltage characteristics of some of the inverted OPVs. At 160 nm, the reduction in PCE is mainly due to the decrease in short-circuit current (J_{sc}); open-circuit voltage (V_{oc}) and fill factor (FF) are nearly unchanged. At 210 nm, J_{sc} is similarly decreased but FF is improved enough that the decrease in J_{sc} is compensated. As a result, PCE increases only slightly. At 430 nm, a decrease in J_{sc} is no longer observed, and the increases in J_{sc} and FF contribute to the improvement in PCE.

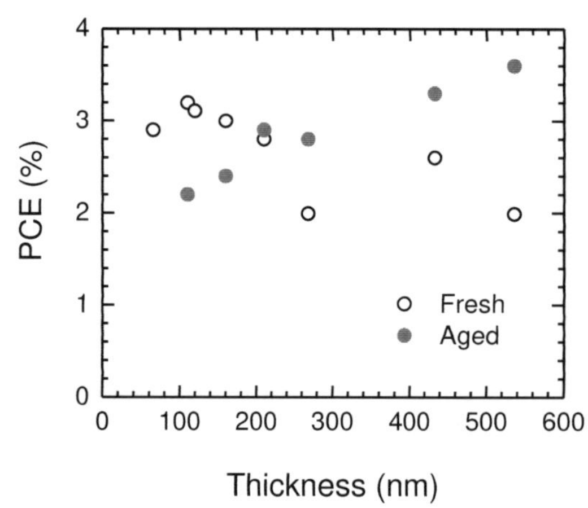

Figure 1. PCEs of inverted OPVs with various active-layer thicknesses. Open (black and white) and filled (red) circles represent PCEs measured immediately after device fabrication and after aging for 2–3 months, respectively.

Degradation mechanisms of OPVs have been intensively investigated [15–20]. The decrease in J_{sc} is caused by chemical degradation of the semiconducting materials or a decrease in the P3HT/PCBM interface area (*i.e.*, an increase in the size of the P3HT or PCBM domains). It has been confirmed that diffusion of fullerenes takes place, even below the glass transition temperature, which leads to acceleration of the phase segregation and to the formation of large domains [15–20]. If the domains become larger than the exciton diffusion length, J_{sc} is expected to be lowered. In our fresh OPVs, the domain size is considered to be smaller than the optimum value. Thus, the increase in domain size with time would result in the development of carrier transport networks from the interface to the electrodes and in the enhancement of carrier drift mobility within the domains; both of these changes would contribute to the enhancement of J_{sc} and FF. On the other hand, chemical degradation of the semiconducting materials may be a reason for the decrease in PCE observed in thin OPVs.

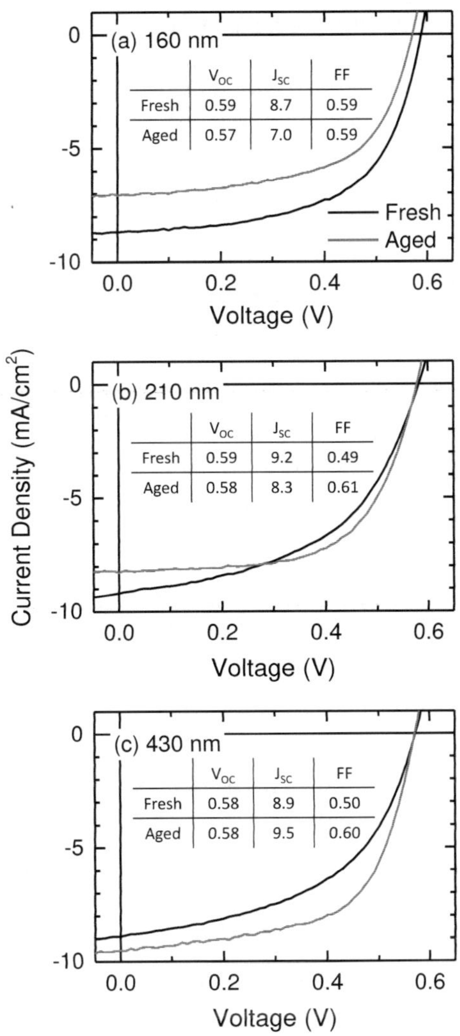

Figure 2. Current–voltage characteristics of inverted OPVs with active-layer thicknesses of (a) 160 nm, (b) 210 nm, and (c) 430 nm. In the insets, we show the typical photovoltaic properties.

IOP Conf. Series: Journal of Physics: Conf. Series **924** (2017) 012009 doi:10.1088/1742-6596/924/1/012009

In addition to the diffusion of fullerene and the chemical degradation, evaporation of residual solvent should be related to the observed change in PCE with time. Although our drying conditions are typical for P3HT:PCBM-based OPVs, there is a possibility that residual solvent reduces PCE, particularly in thicker OPVs, but gradually evaporates with time leading to an increase in PCE. In fact, when thick active layers were dried in a vacuum chamber for longer periods of time (*e.g.* 2 h), J_{sc} was improved. As an example, we show the current–voltage characteristics of inverted OPVs with an active-layer thickness of 530 nm in Fig. 3. Their photovoltaic properties are summarized in Table 1. In the case of 530 nm-thick inverted OPVs, as a result of the improvement of J_{sc}, PCE is increased from 2.0% to 2.6%. If a longer drying time was used for thin OPVs (*e.g.* 100 nm-thick OPVs), their photovoltaic properties were not changed at all. This means that the drying time influences only the amount of residual solvent. It should be noted that no further increase in J_{sc}, FF, or PCE is obtained if thick active layers are dried for longer times. Thus, 2 h seems to be long enough to remove the residual solvent in the 530 nm-thick active layer. The point that we would like to stress here is that unlike J_{sc}, FF is not improved by removing the residual solvent (see Table 1). Compared with the aged OPVs with PCE of 3.6%, the fresh OPV that was dried for 2 h exhibited similar V_{oc} and J_{sc} but lower FF. This result suggests that the improved PCEs of aged, thick OPVs are realized by two factors: the increase in the domain size and the evaporation of the residual solvent.

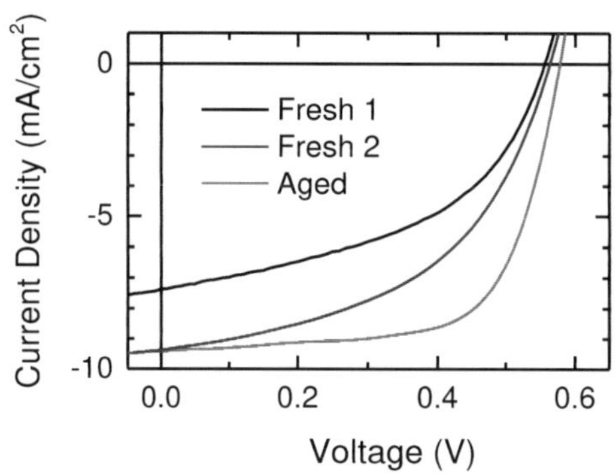

Figure 3. Current-voltage characteristics of inverted OPVs with an active layer thicknesses of 530 nm. Fresh 1 (black) and 2 (blue) represent OPVs dried in a vacuum for 1 and 2 h, respectively. PCEs of those OPVs are 2.0 (Fresh 1), 2.6 (Fresh 2), and 3.6% (Aged). For other photovoltaic properties such as J_{sc} and FF, see Table 1.

Table 1. Photovoltaic properties of inverted OPVs with active-layer thicknesses of 530 nm.

Drying Time (h)	V_{oc} (V)	J_{sc} (mA/cm^2)	FF	PCE (%)
1	0.56	7.4	0.48	2.0
2	0.56	9.4	0.49	2.6
1[a)]	0.58	9.4	0.67	3.6

[a)] Measured 2 months after the device fabrication

As mentioned in the introduction, there is a large variation in the reported optimum thickness for the active layer in OPVs. Inadequate drying in thicker active layers may be one possible reason for the variation. The residual solvent can also be removed during thermal annealing and deposition of metal electrodes. Thus, in some reports, an additional drying process was not carried out. If drying of the active layer is not adequate, FF would decrease with increasing active-layer thickness. In other words, such a tendency might indicate that the drying conditions are not adequate. As shown in Fig. 1, our fresh and aged OPVs exhibit the best PCEs of 3.2% at 110 nm and 3.6% at 530 nm, respectively. These values are close to the reported maximum PCE of P3HT:PCBM-based OPVs (around 4% under simulated 100 mW/cm^2 irradiation) [1, 5, 22, 23]. Therefore, further improvement of the PCE of our inverted OPVs may be limited. Considering this fact, it is expected that the true thickness dependence of the PCE of fully optimized OPVs is much weaker than those shown in Fig. 1 and in some previous reports. On the other hand, Moulé *et al.* reported that the PCEs of their OPVs employing a conventional architecture are almost constant in the thickness range 100–400 nm [5]. Based on our examination, among the various reports on the optimum thickness of P3HT:PCBM-based OPVs, the results of Moulé *et al.* seem to be the most reasonable.

4. Conclusions

In this work, P3HT:PCBM based inverted OPVs with various active layer thickness were fabricated. Fresh OPVs exhibit the best PCE around 110 nm, whereas the optimum thickness becomes more than 530 nm a few months after the device fabrication. We discussed three possible causes of the change in PCE, *i.e.* the residual solvent in particular in thick OPVs, the increase in the domain size of P3HT or PCBM, and the chemical degradation. It is found that the residual solvent mainly reduces J_{sc}. Considering the change in PCE with time, the active layer thickness dependence of PCE of fully optimized OPVs is expected to be much weaker than those observed in this work.

Acknowledgements

This work is supported in part by JSPS KAKENHI Grant Number 15H03883, and by a Grant-in-Aid for Scientific Research on Innovative Areas "New Polymeric Materials Based on Element-Blocks (No. 2401)" (No. JP24102011).

References

[1] Li G, Shrotriya V, Yao Y, Yang Y 2005 *J. Appl. Phys.* **98** 043704
[2] Reyes-Reyes M, Kim K, Dewald J, López-Sandoval R, Avadhanula A, Curran S, Carroll D L 2005 *Org. Lett.* **7** 5749
[3] Sievers D W, Shrotriya V, Yang Y 2006 *J. Appl. Phys.* **100** 114509
[4] Moulé A J, Bonekamp J B, Meerholz K 2006 *J. Appl. Phys.* **100** 094503
[5] Moulé A J, Meerholz K 2008 *Appl. Phys. B* **92** 209
[6] Boland P, Namkoong G 2010 *Jpn. J. Appl. Phys.* **49** 030205
[7] Nam Y M, Huh J, Jo W H 2010 *Sol. Energ. Mat. Sol. Cells.* **94** 1118
[8] Zeng L, Tang C W, Chen S H 2010 *Appl. Phys. Lett.* **97** 053305
[9] Zhao D W, Tan S T, Kec L, Liu P, Kyaw A K K, Sun X W, Lo G Q, Kwong D L 2010 *Sol. Energ. Mat. Sol. Cells.* **94** 985
[10] Jin H, Olkkonen J, Tuomikoski M, Kopola P, Maaninen A, Hast J 2010 *Sol. Energ. Mat. Sol. Cells.* **94** 465
[11] Morvillo P, Bobeico E, Esposito S, Diana R 2012 *Energy Procedia* **31** 69
[12] Kirchartz T, Agostinelli T, Campoy-Quiles M, Gong W, Nelson J 2012 J. Phys. Chem. Lett. **3** 3470
[13] Apaydın D H, Yıldız D E, Cirpan A, Toppare L 2013 *Sol. Energ. Mat. Sol. Cells.* **113** 100
[14] Rahmani R, Karimi H, Ranjbari L, Emadi M, Seyedmahmoudian M, Shafiabady A, Ismail R 2015 *Plasmonics* **10** 495
[15] Yang X, van Duren J K J, Janssen R A J, Michels M A J, Loos J 2004 *Macromolecules* **37**
[16] Hoppe H, Niggemann M, Winder C, Kraut J, Hiesgen R, Hinsch A, Meissner D, Sariciftci N S 2004 *Adv. Funct. Mater.* **14** 1005
[17] Jo J, Kim S S, Na S I, Yu B K, and Kim D Y 2009 *Adv. Funct. Mater.* **19** 866
[18] Watts B, Belcher W J, Thomsen L, Ade H, Dastoor P C 2009 *Macromolecules* **42** 8392
[19] Conings B, Bertho S, Vandewal K, Senes A, D'Haen J, Manca J, Janssen R A J 2010 *Appl.*

Phys. Lett. **96** 163301
[20] Grossiord N, Kroon, J M, Andriessen R, Blom P W M 2012 *Org. Electron.* **13** 432
[21] Li G, Shrotriya V, Huang J, Yao Y, Moriarty T, Emery K, Yang Y 2005 *Nat. Mater.* **4** 864
[22] Huang W Y, Lee C C, Hsieh T L 2009 *Sol. Energ. Mat. Sol. Cells.* **93** 382

IOP Conf. Series: Journal of Physics: Conf. Series **924** (2017) 012010 doi:10.1088/1742-6596/924/1/012010

Functional design of electrolytic biosensor

D M Gamage Preethichandra[1], E M I Mala Ekanayake[1], M Onoda[2]
[1]School of Engineering and Technology, Central Queensland University, Rochhampton Campus, Bruce Hwy Rockhampton, QLD, 4702 Australia
[2]Graduate School of Engineering, University of Hyogo, 2167 Shosha, Himeji, Hyogo, 671-2280 Japan

E-mail: ss-syu02@eng.u-hyogo.ac.jp

Abstract. A novel amperometric biosensbased on conjugated polypyrrole (PPy) deposited on a Pt modified ITO (indium tin oxide) conductive glass substrate and their performances are described. We have presented a method of developing a highly sensitive and low-cost nano-biosensor for blood glucose measurements. The fabrication method proposed decreases the cost of production significantly as the amount of noble metals used is minimized. A nano-corrugated PPy substrate was developed through pulsed electrochemical deposition. The sensitivity achieved was 325 mA/(Mcm^2) and the linear range of the developed sensor was 50-60 mmol/*l*. Then the application of the electrophoresis helps the glucose oxidase (GO_x) on the PPy substrate. The main reason behind this high enzyme loading is the high electric field applied across the sensor surface (working electrode) and the counter electrode where that pushes the nano-scale enzyme particles floating in the phosphate buffer solution towards the substrate. The novel technique used has provided an extremely high sensitivities and very high linear ranges for enzyme (GO_x) and therefore can be concluded that this is a very good technique to load enzyme onto the conducting polymer substrates.

1. Introduction

The commencement of 21st century have witnessed the advanced visual communication-oriented society which is beginning more and more complicated leading to the emergence so-called "ubiquitous" network. In order to meet the demands of such a visual communication-oriented society, it is essential to realize the functional diversity and the functional integration based on the hyperfine processing. However, the techno-scientific innovation in the semiconductor electronics seems to exhibit a downfall owing to the reachable physical limits.

Organic materials are quite fascinating due to availability of many metastable states which is considered to play a pivotal role for the next-generation electronics [1]. Especially, "Organic Iontronics" which are dealing with the devices and applications using functions of ions *etc.* attracts much attention [2-5]. The word "Iontronics" is still unfamiliar words and has been created by the fusion of two words "Ion" and "Electronics". Though it may be said that "Iontrnics" is discipline which deals with phenomena and device applications based on the behavior of the electron and ion, most people may feel like there is something a little off with in comparison with words infiltrating to the general public widely so-called "Electronics". However, the most familiar area biotechnology which is based on amicable use of unique biological functions towards betterment of human life and

[2] To whom any correspondence should be addressed.

environmental conservation cannot deny the important role played by the ions. "Iontronics" does not mean to simply mimic electronics but it is intended to create possible novel and precise features assisted by ions. In order to realize the molecular devices such as functional element of the next-generation electronics, a thought of "Iontronics" will be indispensable.

In this article, the problems that should solve about the characteristic improvement such as the performance of the biosensor which is one of the bio-related "Iontronics" devices are introduced concretely. The biosensor is the measurement device which can be used or imitated by the high material recognition function to have of the bio-material. Since Clark and Lyons proposed the glucose biosensor for the first time in 1962, by putting together with various signal conversion devices and bio-elements, those have been developed and their applications are expected in many fields such as medical care, food, and environment [6].

Figure 1 shows a basic principle of biosensor and their application fields [7]. The biosensor is constructed in the substrate recognition region called the receptor which recognizes a substrate (measurement target material), and in the signal conversion region called the transducer which causes a chemical change at that place and finally converts the change into an electrical signal. In the substrate recognition department, an enzyme, an antibody, a cell, a microbe and so on are used and in the signal conversion region, an electrode detecting substances change, thermistor detecting a heat change, piezoelectric elements detecting a weight change, and so on are employed, The current biosensor is mainly constructed with the enzyme electrode consisting of an enzyme as the substrate recognition region, and the electrode as the signal conversion region.

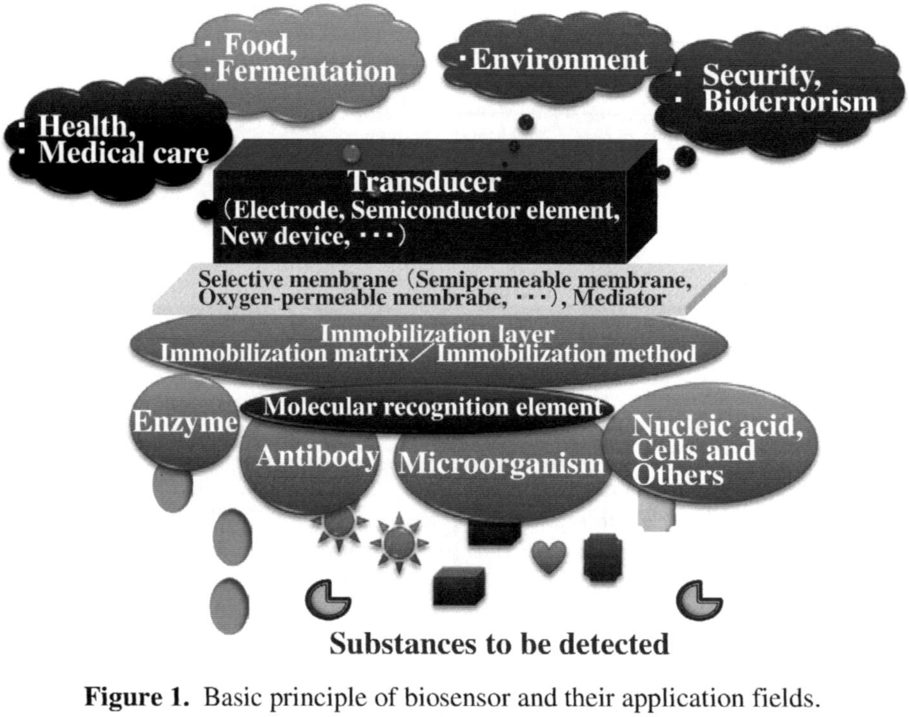

Figure 1. Basic principle of biosensor and their application fields.

The most popular enzyme used in the glucose biosensors is glucose oxidase (GO_x). Various immobilization approaches such as physical adsorption, covalent binding, crosslinking, and coentrapment, have been developed for immobilizing GO_x [8, 9]. As one of effective methods of fabricating glucose biosensors, a combined procedure of physical adsorption and coentrapment of GO_x in conductive polymer matrix may become important [10, 11].

To date, various conductive polymers have been synthesized, but polypyrrole (PPy), the molecular structure of which is shown in Figure 2, is the most stable after doping and is one of the most attractive materials, together with polyaniline (PAn), because of their good environmental stability and relatively high electrical conductivity. Then to develop nerve stimulation electrodes of the inside of the body burial type, PPy of biocompatibility was mainly evaluated in the present study [12].

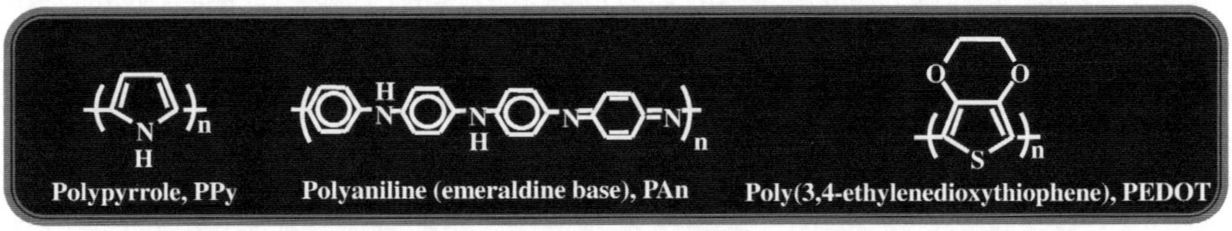

Figure 2. Molecular structures of polypyrrole (PPy), polyaniline (PAn), and poly(3, 4-ethylenedioxy thiophene) (PEDOT).

The sensitivity of a biosensor mainly depends on the amount of enzyme loaded onto the reacting surface and the ability to collect the electrons released by the reaction. Therefore increased effective reacting area is one way to obtain a higher sensitivity. Different research groups apply many different techniques to increase the effective surface area including the use of materials like carbon nanotubes and processes like nano-templating [10, 13-17]. In this research, we have investigated developing a nano-scale corrugated conductive polymer surface suitable for immobilizing glucose oxidase (GO_x) for the development of a low-cost nano-biosensor. That is, traditional biosensors use noble metals for their electrodes and they contribute to a significant portion of their production cost. In contrast, conductive polymer based biosensors use minimum amount of noble metals and therefore reduce the cost of production. Further, the fabrication process involved with this novel nano-biosensor is comparatively simple and low-cost.

2. Electrolytic polymerization technique

2.1. Electrolytic polymerization condition

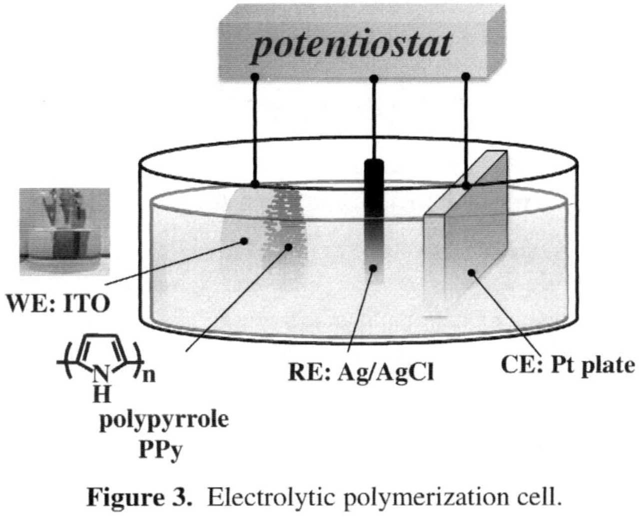

Figure 3. Electrolytic polymerization cell.

In general, electrolytic polymerization technique is performed with the electrolytic polymerization equipment shown in Figure 3 [18]. If the aromatization compound monomer that is to be polymerized is dissolved into a solvent containing a suitable supporting electrolyte and an appropriate voltage is applied to the electrode pair immersed into this solution, the monomer is oxidized or reduced on the surface of the anode or cathode, respectively, and it polymerizes in the form of a powder or a film and at times may show arborization. The case in which the monomer is oxidized and polymerized on the anode surface is called electrolytic oxidation polymerization.

On the other hand, the case in which the monomer is reduced and polymerized on the surface of the cathode is called electrolytic reduction polymerization. The reference electrode might be immersed in the solution if necessary.

In order to obtain good-quality conductive polymer films by the electrolytic polymerization technique, it is necessary to examine the above predominant factors in detail and to understand the optimal electrolytic polymerization condition. Moreover, the conductive polymer can be reversibly electrochemically doped and dedoped and the dopant concentration can be widely adjusted by limiting the voltage, enabling a film that has an arbitrary electric conductivity to be obtained.

2.2. Reaction mechanism of electrolytic polymerization

The reaction mechanism of conductive polymer synthesis by electrolytic polymerization has not been clarified, because various factors, such as the composition of the electrolyte and the electrolytic conditions, interact very complexly with the electrode reaction. Therefore, the polymerization reaction condition differs for each individual conductive polymer and the optimum conditions are determined through trial and error. However, qualitatively, the following reaction mechanisms are accepted in general. The dynamic movement of the molecule, which takes part in the monomer, and the electrolytic ion formed by the transfer of an electron from the solvent are caused at the interface between the electrode and the electrolyte in the electrolytic polymerization technique. Usually, 2–2.5 electrons are consumed in the electrolytic polymerization reaction, and ultimately, two pieces of electrons are used in the polymerization reaction and the remainder is used in the doping. As a result, the same number of protons (H^+) that as the number of electrons used for the polymerization will accumulate in the polymerization liquid. Therefore, the polymerization reaction occurs as a result of the removal of the electron from the monomer, and the coupling reaction, in which the generated radical cation (positive ion) is assumed to be the activated species, and the deprotonation reaction are thought to progress repeatedly. The polymerization reaction is thought to be either reaction 1, the parent electron substitution coupling reaction, or reaction 2, the radical coupling reaction (Here, M indicates a monomer), as shown in Figure 4.

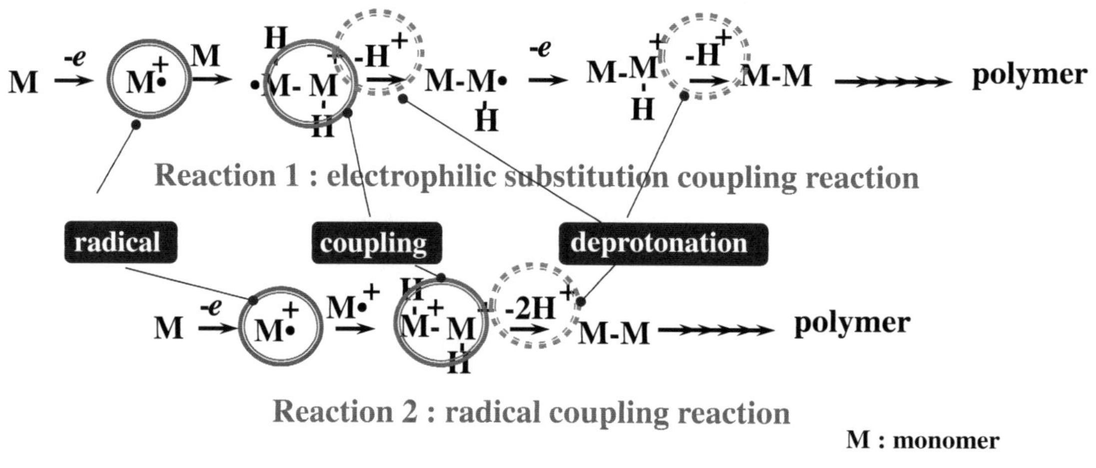

Figure 4. Reaction mechanism of electrolytic polymerization.

However, the structure analysis of the obtained polymer is difficult because the polymer is nonfusible and nonsoluble. Then, the polymerization reaction mechanism itself becomes very complex and a united opinion is not obtained, because electrolytic polymerization is a reaction of a nonuniform system that progresses in a limited area in the neighborhood of the electrode, and it is necessary to consider the supply of monomers and electrolytic ions to that area. That is, the electrolytic polymerization is thought to begin with the oxidizing or reducing reaction of the monomer in the electrolyte upon applying potential, followed by the generation of the radical cation or the radical anion in the aromatic compounds, and subsequently, the polymerization progresses repeated the coupling and the deprotonation reaction. It is necessary to prevent the solvent from undergoing the electrolytic reaction with the monomer at a low voltage, because the effects of the various factors such as the type of solvent, the supporting electrolyte, the polymerization voltage, and the polymerization temperature, on the electrolytic polymerization reaction are not yet fully clarified [18].

3. Experimental
Polymerizing a nano-corrugated conductive polymer substrate matching the size of the enzyme size is not an easy task. In the past we have done this through nano-templating using a nano-porous almina template [10, 11]. However the cost involved with the nano-porous alumina template was a significant factor in the cost of production. Therefore in this research we try to develop a different technology to produce a matching conductive polymer surface for the enzyme immobilizing.

3.1. Materials
Glucose oxidase (E.C.1.1.3.4., 210 U/mg, GO_x) from Aspergillus niger, pyrrole monomer, ascorbic acid, uric acid and sodium tetrafluorophosphate were purchased from Wako, Japan. α-D glucose was obtained from Sigma-Aldrich. Pyrrole monomer was distilled and sodium tetrafluorophosphate solution was freshly prepared before use. A 0.05 mol/*l* phosphate buffer solution (PBS) was freshly prepared using Potassium Dihydrogenphosphate (KH_2PO_4) and Disodium Hydrogenphosphate (Na_2HPO_4) at pH 6.5. The pH values of the buffer solutions were adjusted using a Cyberscan pH 100/RS232 portable meter. A stock solution of 1 mol/*l* glucose was prepared and allowed anomers to reach equilibrium at aqueous state, before use. That is, glucose solutions made from α-D glucose (prepared with a 0.05 mol/*l* PBS, pH 6.5) were left at rest for more than 12 hours before use and allowed to stabilize. All the chemicals were of analytical grade and throughout the experiment temperature was maintained at 25 ± 2 °C unless otherwise stated.
ITO (indium tin oxide) electrodes were Pt coated by plasma sputtering using JEOL quick coater (JFC 1,500). Hokuto Denko automatically polarization system (HSV-100) was used to perform all electrochemical studies including polymerization and current response measurement.

3.2. Electrolytic polymerization equipment

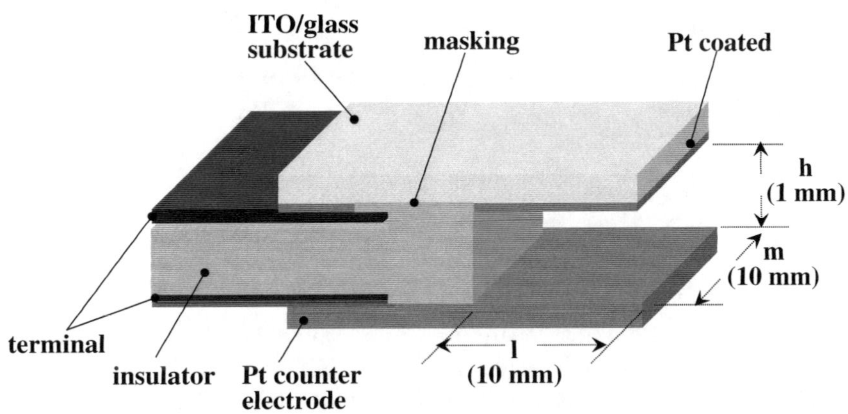

Figure 5. Schematic diagram of the sensor fabrication arrangement.

In general, as shown in Figure 3, a conventional three-electrode cell consisting of a working electrode (Pt coated ITO), a platinum counter electrode and an Ag/AgCl reference electrode was used for

IOP Conf. Series: Journal of Physics: Conf. Series **924** (2017) 012010 doi:10.1088/1742-6596/924/1/012010

electrochemical synthesis and measurements. A magnetic stirrer was used during the amperometric measurements to ensure convective transport. However, the schematic experimental setup we used for our work is shown in Figure 5. The ITO electrode was modified by plasma vacuum deposition of a 50 nm thick Pt layer on the ITO surface using JEOL quick auto coater. This electrode was placed on one contact terminal of the test ring and a Pt electrode was placed on the other contact terminal where the Pt/ITO surface facing the Pt counter electrode. The gap between them was 1 mm and the common area between them was 1 cm^2. The test rig was positioned to have the electrodes horizontal and an adequate amount of 0.05 mol/*l* aqueous pyrrole solution doped with 0.1 mol/*l* $NaPF_6$ was carefully injected in between the electrodes.

3.3. Fabrication of nano-porous PPy electrodes

Normally, the sensitivity of a biosensor is determined by several factors such as the match between surface porosity and physical dimensions of target enzyme, effective surface area and charge collecting backbone material on the substrate *etc*. Therefore, it is very important to increase the effective surface are in order to obtain a good sensor. However, the surface roughness of the artificial porosity should match with the physical dimensions of the target enzyme [9].

When a PPy layer is electrodeposited on a planar electrode, polymerizing starts with random nuclei and keeps on growing out of them as long as the polymerization current continues. However, if the polymerization current is interrupted instantaneously, there is no guarantee that it will restart from the same point it stopped before and most probably start from a different physical location on the working electrode where there is a low contact resistance. We have effectively utilized this random polymerization phenomenon to develop an artificially porous conductive polymer surface for our working electrode to entrap GO_x. The physical dimensions of the porosity can be controlled with the applied voltage, mark to space ratio and pulse width of the polymerization potential. In this research we have tested various pulse timing and mark to space ratios and optimized our sensor performance with 1 s pulse time with 1 : 1 mark to space ratio of 1 V polymerization potential. Figure 6 shows a schematic experimental diagram of the pulsed electrolytic polymerization system.

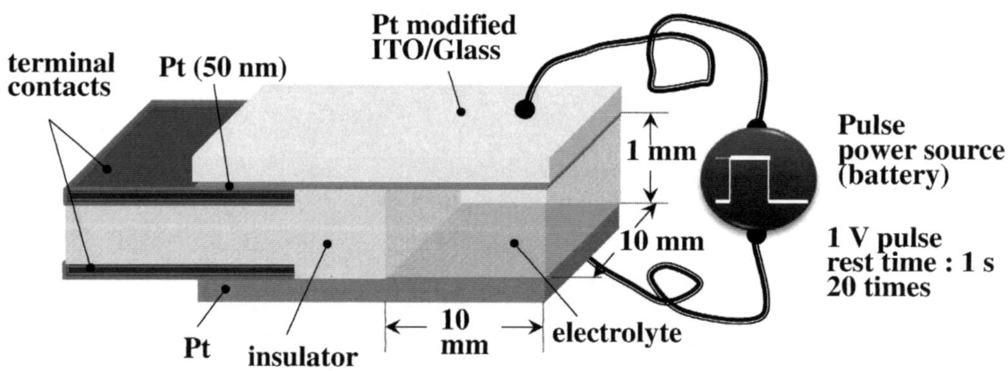

Figure 6. Schematic experimental diagram of the pulsed electrolytic polymerization system.

3.4. Surface morphology of PPy electrodes

Different pulse timing and pulse voltages were used to optimize the above parameters. Figure 7 (a) and (b) shows the nano-scale morphological changes of the PPy substrates electrodeposited under pulsed voltage and constant voltage respectively from these trials. Pulse voltage was 1 V and deposition was carried out for 1 s and turned off for the next 1 s period. This was repeated for 175 cycles. Pt modified ITO substrates were used for all experiments. As shown in Figure 7 (a) and (b), clearly, these surface morphologies of these PPy electrodes are greatly different each other. That is, the surface area of pulsed deposition PPy electrode (a) is larger than that of constant voltage deposited PPy electrode (b).

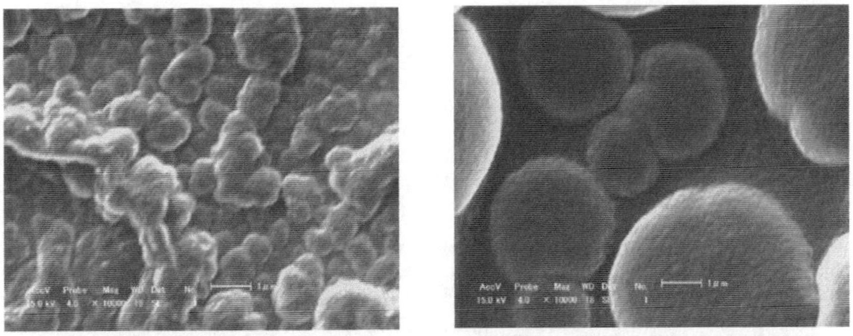

(a) pulse-deposited PPy **(b) DC voltage deposited PPy**

Figure 7. Photographs of scanning electron microscope of polypyrrole (PPy) electrodes
(a) Pulse voltage was 1 V and deposition was carried out for 1 second and turned off for the next one second period. This was repeated for 175 cycles.
(b) DC voltage was 1 V and deposition was carried out for 3 minutes.

3.5. Immobilization of GO_x on the PPy electrode

Once $NaPF_6$ doped pyrrole monomer is in place, twenty times for 1 V pulses of 1 s were applied to the terminals where the Pt/ITO acted as the working electrode and Pt as the counter electrode. After polymerization the remaining pyrrole monomer was removed and both the electrodes were carefully washed in pH6.5 phosphate buffer solutions (PBS). Then the electrode assembly was positioned again to have the electrodes horizontally and a little less than adequate amount of GO_x solution was injected in between the electrodes.

An aliquot of glutaraldehyde was injected to the same chamber to mix with GO_x and a potential of 1 V was applied to the system for 20 minutes constantly. This provides the environment for the GO_x to get deposited under a high electric field of 1 kV/m. This electrophoretic deposition helps the GO_x deposition on the PPy electrode. After the period the sensor and the counter electrode were carefully washed in pH6.5 PBS.

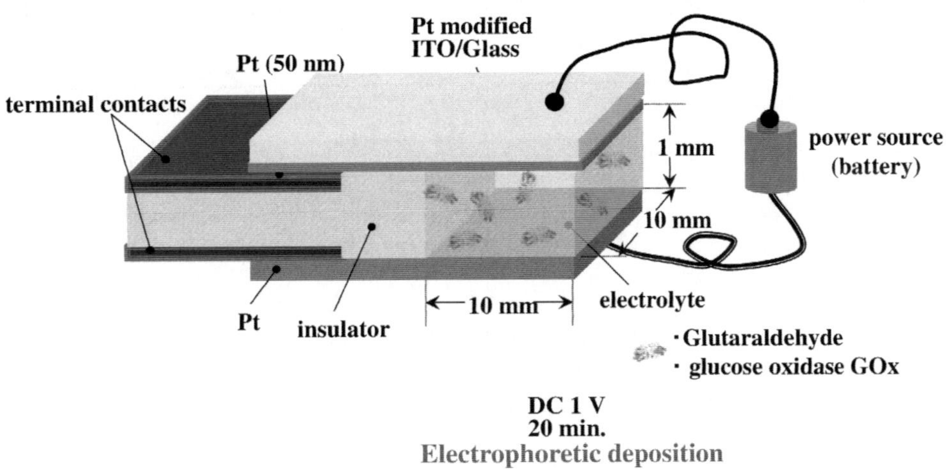

Figure 8. Immobilization of GO_x on the PPy electrode and electrophoretic electrodeposition.

3.6. Biosensor fabrication

The sensor assembly was mounted vertically for the testing where the open end of the electrodes facing downwards. The sensor electrode and counter electrode were gently submersed in a slowly steady stirring PBS of pH6.5 up until the 1 cm×1 cm area was covered by PBS. Fabricated sensor (as working electrode) and the Pt counter electrodes were connected to Hokuto Denko HSV-100 electrochemical system in a three-electrode configuration where the reference electrode was Ag/AgCl. The GO_x/pulse-deposited PPy/Pt electrode was maintained at room temperature at a polarization potential of 0.38 V vs. Ag/AgCl in the air-saturated sensing solution to yield a stable background current. The system was applied a potential of 0.38 V and was allowed to settle down before the testing started. When the system reached stability a known concentration of glucose was added to the system and allowed to settle. This was repeated for different glucose concentration and all data were logged.

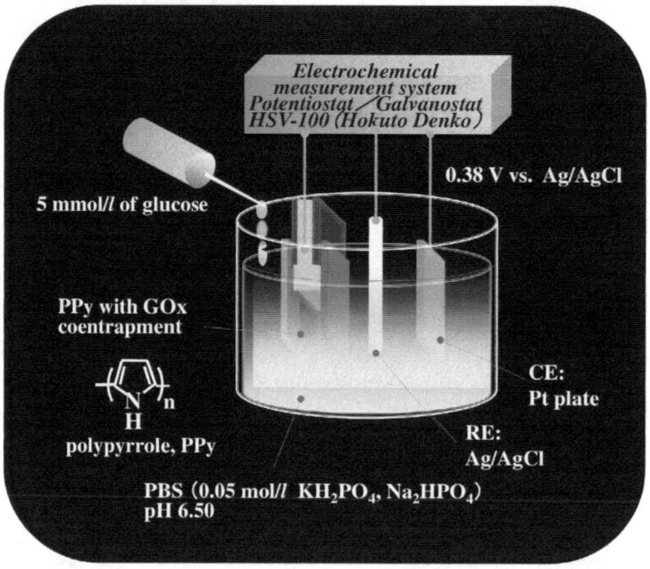

Figure 9. Experimental systems of the biosensor performance evaluation.

In the next round of testing the system was applied a potential of 0.38 V and was allowed to settle down before the testing started. When the system reached stability, 5 mmol/*l* of glucose was added to the solution at constant time intervals and the response of the system was continuously recorded through the HSV-1000.

The sensitivity of a sensing electrode is represented by the following.

$$sensitivity = \left(current\ response\right) \Big/ \left(concentration\ of\ glu\cos e\right) \times \left(initial\ area\ of\ the\ electrode\right)$$

All sensors, PBS, and GO_x were kept at 4 °C while in storage and were allowed to settle to the room temperature before the experiments.

3.7. Equipment which assists experiments

Electrochemical impedance spectroscopy (EIS) measurements system (Hokuto Denko Co., Ltd., HZ-5000) was used to confirm the surface modification under GO_x deposition stage. That is, two EIS measurements, one before and one after GO_x deposition were carried out to make sure an effective GO_x deposition was taken place.

If the enzyme was loaded on to the PPy electrode, it should modify the nanoscale morphology of the sensor substrate and hence the contact impedance. The changes in real and imaginary impedance show that the enzyme deposition has taken place successfully. However, EIS measurement in our experiment is only adhesion confirmation of GO_x onto the PPy electrode.

The details of the EIS measurement based on the electrophoretic electrodeposition of GO_x will be reported somewhere in the near future.

4. Results and discussion

This research revealed that the developed novel glucose sensor has comparatively a high sensitivity and a very good linear range for this class of a nano-biosensor. This was achieved through the novel technique of electrophoretic enzyme deposition. Same sensor substrate deposited with glucose oxidase, again with the same electrophoretic enzyme deposition resulted in a very high sensitivity and a good linear range [19]. Both these tests revealed that the new enzyme deposition technique has a very high impact on the amount of enzyme loaded on to a unit linear surface area.

4.1. Optimization of biosensor's sensitivity

Our purpose of using nano porous corrugated PPy as our base electrode was, to retain its original nano porosity until PPy polymerization stage allowing PPy film to be porous artificially. Therefore controlling the thickness of PPy by optimizing the polymerization time (i.e., application time of the pulsed voltage, in other word, the charge passed during polymerization) was crucial in this regard. With GO_x enzyme having particles around 10-100 nm in size, adequate size of pores for GO_x entrapment was also regarded to be in a closer range.

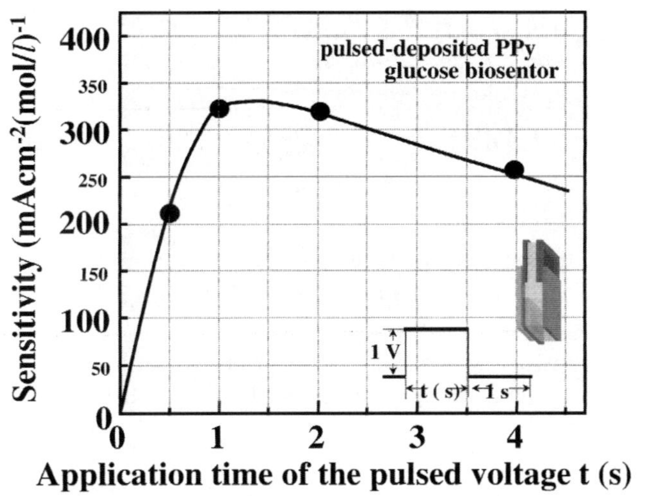

Figure 10. Relation between sensitivity and PPy polymerization time.
(PPy was polymerized at a pulsed voltage of 1 V with a different application time of pulsed voltage.)

Figure 10 shows good performances around a polymerizing time (application time of the pulsed voltage t (s)) of around 1 s giving evidence for ideal pore size to serve our purpose. The SEM image in Figure 7 provides clear clues of this PPy growth. The efficiency of immobilization of the enzyme is influenced by this changed effective surface area available due to nano-pores.

4.2. Time response of the sensor for different glucose concentrations

The sensor was tested for its time response under known different glucose concentration. It displayed very similar and consistent time responses for different glucose concentrations. Results revealed that the sensor reaches its stable state in about 40 s, which is little longer than normal value for a glucose biosensor but still acceptable for a low-cost biosensor.

The current-time curves shown in Figure 11 shows that the GO_x/pulse-deposited PPy/Pt electrodes have a response and high sensitivity to glucose. At a polarization potential of 0.38 V vs. Ag/AgCl and in an air-saturated sensing solution, the anodic current increased dramatically and reached a steady state within 40 s as seen the inset in Figure 11 after the injection of a certain amount of a glucose solution. By applying a polarization potential of 0.38 V, polarization current, that is, background current was reached a steady state. And then, only the response current after the injection of a certain amount of glucose solution was examined.

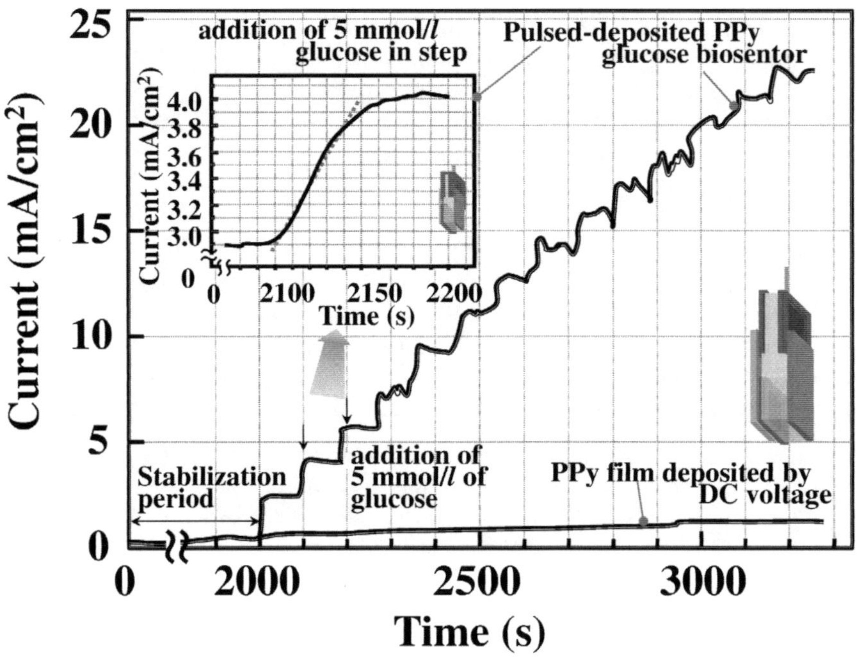

Figure 11. Response of the pulse-deposited PPy glucose biosensor against addition of 5 mmol/*l* glucose in steps.

Figure 11 shows the response of the fabricated sensor against increase of glucose concentration by 5 mmol/*l* at a step. The system was allowed to settle for extensive amount of time in order to minimize background noise from the PBS, but it can be used with much less stabilization if appropriate noise filtering is employed.

The current response of a PPy film polymerized by constant voltage on a Pt coated ITO electrode is shown in Figure 11 for similar concentration of glucose. It has been revealed that only a little amount of GO_x is entrapped in the PPy film deposited by constant voltage on the Pt/ITO electrode without a porous structure compared to the highly porous new pulse-deposited PPy electrode.

The results illustrated in Figure 11 reveal that the sensor has a very high sensitivity and a high linear range. Figure 12 was derived from data in Figure 11 and this shows that the linear range spans from 0 to 60 mmol/*l*, which is more than double the range of a blood glucose sensor. Further, the sensitivity value calculated from Figure 11 for this sensor is 325 mA/(Mcm^2). This is a very high sensitivity enables designing of low sensitivity front-end electronics otherwise needed very high sensitivity measuring circuits [21, 22].

As it can be seen from the Figure 11, the response current tends to be fluctuating at higher glucose concentrations. This can be due to the physical arrangement of the sensor electrodes with only 1 mm apart from each other, not allowing a homogeneous mixture of glucose to enter the nature gap at higher concentrations. At the same time, the reaction product is not removed at a linear rate due to narrow opening and surface tension of the liquid. Both these factors contribute at higher glucose concentrations to produce a fluctuating current. This can be overcome by introducing a planar geometry, but will have to scarify the sensitivity at the same time.

Though the sensor responded for even higher glucose concentrations, the sensitivity has a higher standard deviation beyond 55 mmol/*l* and the linear range ends nearly at 60 mmol/*l* of glucose concentration. The extraordinary sensitivity and the very high linear range resulted by the pulsed deposition of PPy film suggests that this newly developed biosensor can be applied in industrial applications as well.

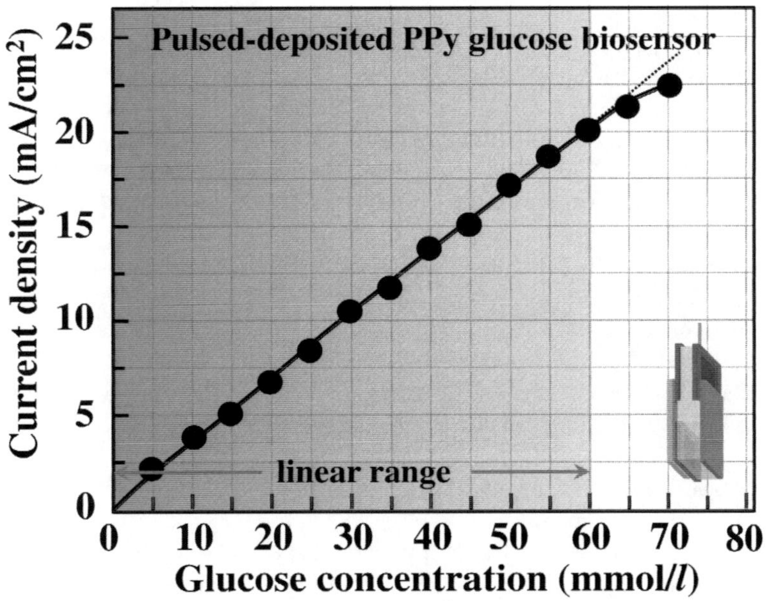

Figure 12. Sensitivity of the developed nano-biosensor against glucose concentrations.

5. Conclusion

The commencement of 21st century have witnessed the advanced visual communication-oriented society which is beginning more and more complicated leading to the emergence so-called "ubiquitous" network. In order to meet the demands of such a visual communication-oriented society, it is essential to realize the functional diversity and the functional integration based on the hyperfine processing. However, the techno-scientific innovation in the semiconductor electronics seems to exhibit a downfall owing to the reachable physical limits. Organic materials are quite fascinating due to availability of many metastable states which is considered to play a pivotal role for the next-generation electronics.

In this research, highly sensitive glucose oxidase (GO_x) electrodes were fabricated on the basis of nanostructured polypyrrole (PPy) films using pulse electrodeposition. The nanostructures of the PPy films had a nano-porous morphology like voids. GO_x was immobilized in nanostructured PPy films coated on a ITO substrate electrode. The GO_x/PPy/Pt electrode showed a linear response of glucose concentration in the range of 0 – 50 or 60 mmol/*l* at a potential of 0.38 V (vs. Ag/AgCl). Its sensitivity was measured to be approximately 325 mA/(Mcm^2) at room temperature. In comparison, the sensitivity of the GO_x/PPy/Pt electrode based on a flat PPy film was only approximately 0.3 mA/(Mcm^2) under the same conditions.

The nano-biosensor fabricated shows excellent performance with the optimized polymerization parameters. These results show a cost effective way of developing a highly effective nanostructured working electrode surface for nano-biosensor application. This method can be used to develop different nano-biosensor working electrodes by fine-tuning the electrochemical deposition parameters such as time interval, mark to space ratio and pulse shape *etc.*, to adjust the PPy cavity sizes to suit the chosen enzyme.

The electrochemical deposition of pyrrole under pulsed electric field has created many different PPy islands on the planar Pt modified ITO surface otherwise it would have been a linear growth from the original nuclei. Eventually this produced an artificially corrugated nanostructure for the newly formed PPy substrate. From these results, it can be concluded that the PPy layer electrodeposited on the planar surface of Pt modified ITO has got suitable size of nano-cavities to hold the GO_x enzyme molecules.

Reference

[1] Stavrinidou E Gabrielsson R Gomez E Crispin X Nilsson O Simon D T and Berggren M 2015 *Science Advances* **1**(10) e1501136
[2] Leger J Berggren M and Carter S 2011 *Iontronics* (New York, CRC Press)
[3] Iwasa Y and Shimotani H 2015 *OYO BUTURI* **84**(4) 306-in Japanese
[4] Chun H and Chung T D 2015 *Annual Review of Analytical Chemistry* **8** 441-462
[5] Onoda M Kaneto K Ohsawa T and Yoshino K 2016 *Organic Iontronics* (Tokyo, Morikita Publishing Co., Ltd.) in Japanese
[6] Clark L C Jr and Lyons C 1962 *Annals of the New York Academy of Sciences* **102** 29-45
[7] Tanaka A and Matsuno R 1995 *Introduction of Enzyme Engineering* (Tokyo, Corona Publishing Co., Ltd.) in Japanese
[8] Sung W J and Bae Y H 2000 *Analytical Chemistry* **72** (9) 2177-2181
[9] Gerard M Chaubey A and Malhotra B.D 2002 *Biosensors and Bioelectronics* **17** (5) 345-359
[10] Ekanayake E M I M Preethichandra D M G and Kaneto K 2007 *Biosensors and Bioelectronics* **23**(1) 107-113
[11] Preethichandra D M G and Onoda M 2015 *IEEJ Transactions on Electronics, Information and Systems* **135** (2) 174-180 in Japanese
[12] Onoda M Abe Y Tada K Kawakita Y Fujisato T and Uto S 2012 *IEEJ Transactions on Electronics, Information and Systems* **132** (9) 1422-1428 in Japanese
[13] Wang J and Musameh M 2005 *Analytical Chemica Acta* **539** (1-2) 209-213
[14] Gade V K Shirale D J Gaikwad P D Savale P A Kakde K P Kharat H J and Shirsat M D 2006 *Reactive and Functional Polymers* **66** (12) 1420-1428
[15] Zhu L Yang R Zhai J and Tian C 2007 *Biosensors and Bioelectronics* **23** (4) 528-535
[16] Gao Q Gao Y Zhang W Qi H and Zhang C 2011 *Sensors and Actuators B: Chemical* **153** (1) 211-225
[17] Gangadharan R Anabdan V Zhang A Drwiega J C and Zhang G 2011 *Sensors and Actuators B: Chemical* **160** (1) 991-998
[18] Onoda M. Supervisor 2012 *Physical Properties and Evaluation of Conductive Polymers for High Performance Organic Electronic Devices* (Tokyo, CMC Publishing Co., Ltd.) in Japanese
[19] Preethichandra D M G Ekanayake E M I M and Onoda M 2016 Electrophoretic enzyme deposition on nano-structured conducting polymer substrates for high sensitibity *26th Anniversary World Congress on Biosensors Biosensors 2016* (25-27 May 2016, Göthengurg, Sweden. BIOS2016_0806)
[20] Preethichandra D M G and Onoda M 2015 *IEEJ Transactions on Fundamentals and Materials* **136** (11) 697-703 in Japanese
[21] Preethichandra D M G and Shida K 2001 *IEEE Transactions on Instrumentation and Measurement* **50** (6) 1583-1586
[22] Linares-Barranco B and Serrano-Gotarredona T 2003 *IEEE Journal of Solid-State Circuits* **38** (8) 1353-1363

IOP Conf. Series: Journal of Physics: Conf. Series **924** (2017) 012011 doi:10.1088/1742-6596/924/1/012011

Calculation of Reflection Spectrum with Actual Layer Thickness Profile in Nacre of Akoya Pearl Oyster

R Ozaki[1], A Yoshimoto[1], G Watanabe[1], K Kadowaki[1], and K Odawara[2]

[1]Department of Electrical and Electronic Engineering and Computer Science, Graduate School of Science and Engineering, Ehime University, 3 Bunkyo-Cho, Matsuyama, Ehime 790-8577, Japan
[2]Fisheries Research Center, Ehime Research Institute of Agriculture, Forestry and Fisheries, 5516 Shimonada, Uwajima, Ehime 798-0104, Japan

E-mail: ozaki.ryotaro.mx@ehime-u.ac.jp

Abstract. Pearls have beautiful color and luster which are caused by the interference of light within a multilayer in nacre. The nacre is composed of submicron aragonite layers and nanoscale organic sheets of chitin. The value of a pearl is determined by its size, color, and luster. To improve the quality of cultured pearls, we investigate reflection properties and layer thickness profiles in pieces of nacre of Akoya pearl oyster. SEM images show us how large the thickness variation of the nacre of Akoya oyster is. We discuss the reflection properties of the pieces of nacre to connect the relation between the reflection spectra and the thickness profiles obtained by SEM observation.

1. Introduction

People have been fascinated by the beautiful colors of pearl and opal jewelries since a long time ago. The colors of the pearls and opals are the structural colors caused by multiple reflections from a nanostructure [1-3]. The study of the structural color is intimately associated with photonic crystals which have been widely studied to develop new optical applications such as low threshold lasers [4-10] and nanoscale optical waveguides [11,12]. The optical properties of nanostructures in nature can also be analyzed by calculation techniques for the photonic crystals. Not only pearls and opals, but also morpho butterflies and gold bugs have attracted much attention from a scientific point of view [13].

We have investigated the layer structure of Akoya pearl oysters and its structural colors to improve the quality of the pearls harvested from pearl farming [14]. In general, the price of a pearl is determined by its size, color, and luster. For example, pink Akoya pearls are particularly popular with women. However, the color of a pearl cannot be completely controlled in pearl farming. Therefore, a screening method to obtain desired color in pearl farming is very important [14,15]. In addition, the development of grading of a pearl's quality based on optical calculations is also a goal.

A cultured Akoya pearl consists of a polished sphere nucleus made from freshwater mussel shell and a thin coating of nacre. The nacre has a layered structure of aragonite crystals and conchiolins. The aragonite is one of the two common naturally occurring crystal forms of calcium carbonate ($CaCO_3$), and the conchiolin is a complex protein. In general, the thickness range of the aragonite layer is from 300 nm to 700 nm, and that of the conchiolin is from 10 nm to 20 nm. In the point of quality of pearls, it is needless to say that the uniformity of thickness is a key factor. However, the relationship of the layer profile and optical properties of nacre are not investigated in detail yet. In this paper, we observe the layer profiles of several samples to know how much the aragonite layer varies in the nacre and measure their reflection spectra to investigate optical properties. Furthermore, we also try to reproduce reflection spectra of the samples by transfer matrix method using the measured layer profiles.

2. Reflection spectra of pieces of nacre

At the start of pearl farming, a small piece of mantle tissue is cut from an Akoya pearl oyster (piece oyster), and then nucleus with the piece of mantle tissue is inserted into another Akoya oyster (mother oyster), as shown in figure 1. The small piece of mantle tissue from the piece oyster grows up in the mother oyster, and then the piece gradually covers the nucleus. During the process of holding mother oysters in a farming point, the mother oyster provides nourishment to the piece in sea. Therefore, the genetic information of the piece oyster is important because the small piece becomes the nacre of cultured pearl. Here, we focus on the aragonite layer thickness of the piece oyster on the assumption that the color of a cultured pearl is reflected by the layer thickness.

Figures 2(a)-2(d) show the samples of pieces of nacre measured in this study. We prepared the two red pieces of nacre A and B and two pale green pieces of nacre C and D whose sizes were about 1 cm^2. These pieces were cut from red and green Akoya pearl oysters, as shown in figure 2(e). To investigate the optical properties of these pieces of nacre, we measured reflection spectra of the samples. Figure 3 shows the reflection spectra of the pieces of nacre. The reflection spectra of the pieces of nacre were obtained by a charge-coupled device (CCD) multichannel spectrometer (BlueWave-VIS, StellarNet) attached to a microscope (Eclipse 50i POL, Nikon). The reflection spectra of nacre A and B have two peaks at 480 nm and 700 nm. The colors of reflection peaks of 480 and 700 nm correspond to blue-green and deep red, respectively. The peaks are caused by Bragg reflection from the periodic structure

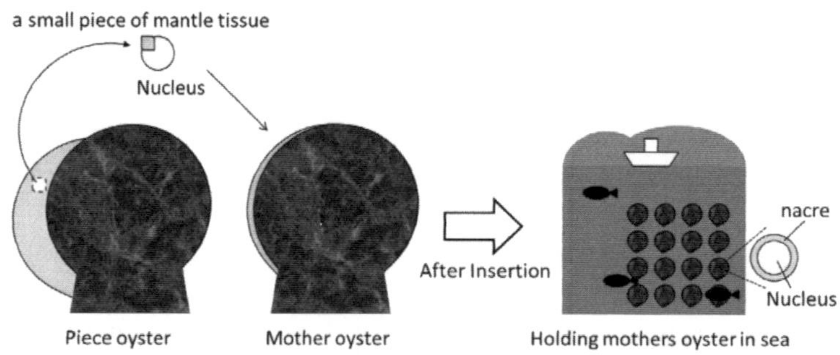

Figure 1. Schematic illustration of process of pearl farming.

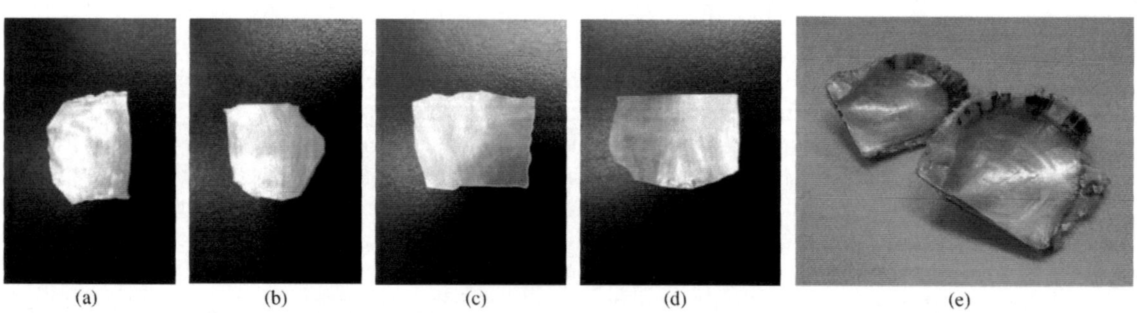

Figure 2. Samples used in this study: (a) nacre A, (b) nacre B, (c) nacre C, (d) nacre D, and (e) green and red Akoya pearl oysters.

IOP Conf. Series: Journal of Physics: Conf. Series **924** (2017) 012011 doi:10.1088/1742-6596/924/1/012011

of the aragonite and conchiolin layers. The lowest point of reflectance is about 40 % because light traveling in the nacre is strongly scattered by defects in the nacre. Due to the strong scattering, the color of the piece of nacre becomes whitish. Therefore, the colors of the pieces of nacre A and B would be slightly pink. Figure 3(c) shows the reflection spectrum of the piece of nacre C. The spectrum has a small peak at 540 nm and a large peak at 820 nm. The piece of nacre C also strongly scattered light, therefore, the base line of the reflection spectrum was about 40 %. The wavelengths of 540 nm and 820 nm correspond to green and near infrared. The small green reflection and light scattering from the piece of nacre C show us the color of pale green, as shown in figure 2(c). Figure 3(d) shows the reflection spectrum of the piece of nacre D. The spectrum does not show a clear large peak, and the reflection at near 520 nm is slightly high. The reflection peak and color of the piece of nacre D is the most unclear in the four samples.

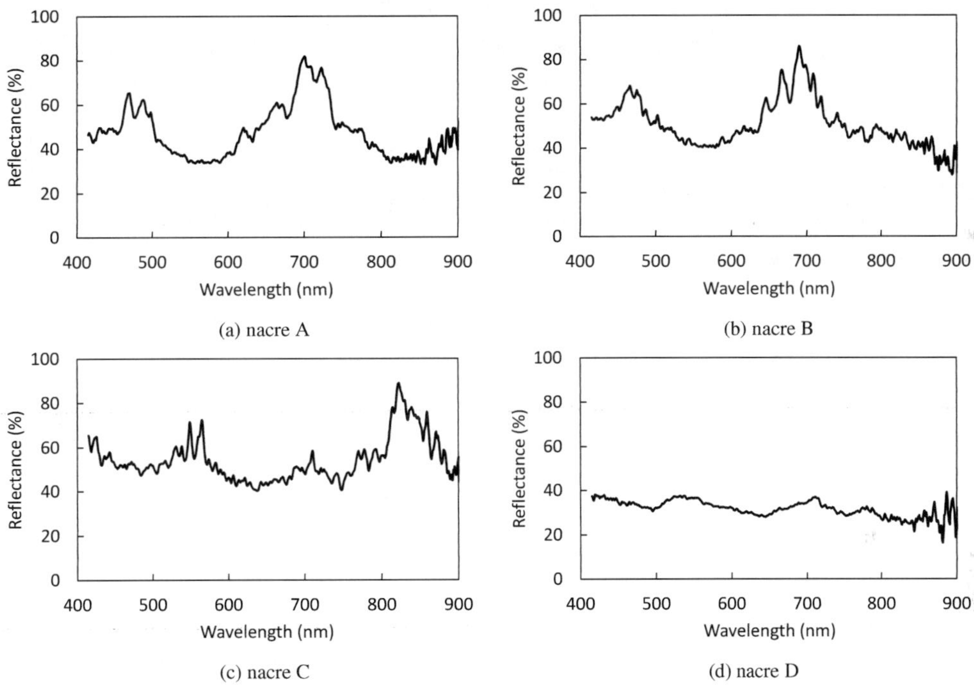

Figure 3. Reflection spectra of the pieces of nacre of Akoya oysters.

3. SEM observation of nacre in Akoya oyster

We observed periodic structures in the nacre of the samples by using scanning electron microscope (SEM) (S-3100H, Hitachi). To investigate the thickness profile of aragonite crystal layers, we observed the four samples from the bottom part to the top part of the pieces. Figure 4 shows the SEM images of the piece of nacre B from the top to the bottom. We arranged 42 SEM images in line to show the thickness profile of the piece. The upper left corner is the top part of the piece of nacre and the lower right corner is the bottom. As mentioned above, the thickness range of the aragonite layer is generally from 300 nm to 700 nm, and that of the conchiolin is from 10 nm to 20 nm. We could not distinguish between thin conchiolin layers and aragonite crystal layers from the SEM images because the SEM did not have enough resolution for observation of a thin conchiolin layer. However, the periodicity of the nacre B could be measured from these images. To obtain the thickness profile in the piece, we counted the number of aragonite crystal layers in all SEM images, and then we estimated the average layer thickness in each SEM image. The estimated layer thicknesses for all pieces are plotted

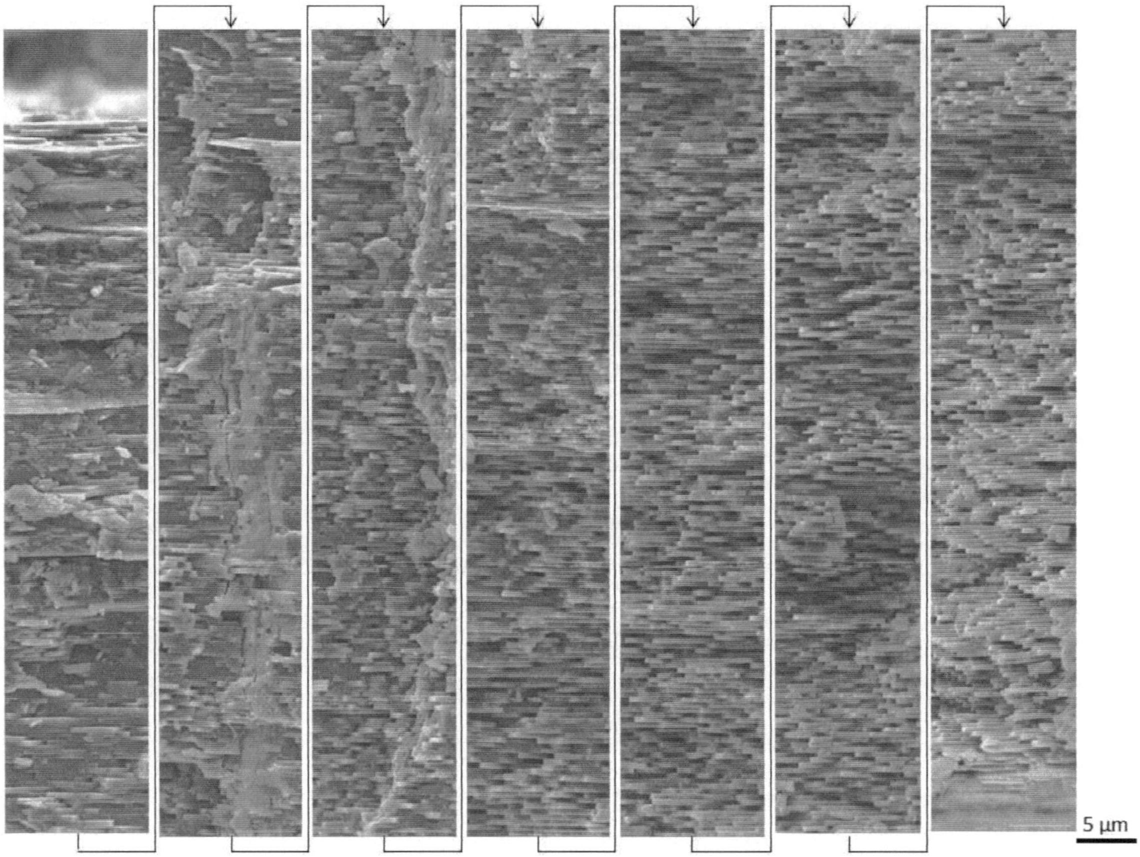

Figure 4. SEM images of the piece of nacre B.

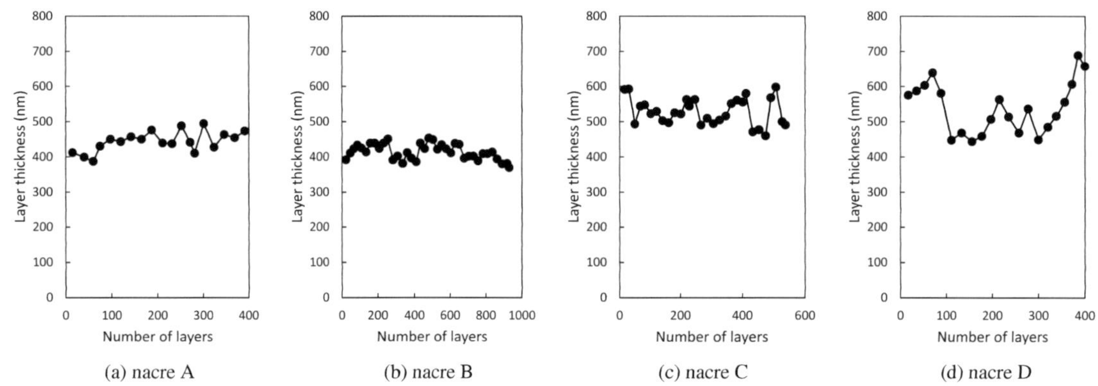

Figure 5. Layer thickness profiles of the pieces of nacre.

in figure 5. Since we just counted the number of layers, this estimated layer thickness corresponds to the total thickness of the aragonite crystal and conchiolin layers. Figures 5(a) and 5(b) show the thickness profiles of the pieces of nacre A and B, respectively. The layer thicknesses for nacre A and B are distributed at around 420 nm, and the thickness profile does not show a large variation through the top to the bottom. In contrast, the thickness profiles of nacre C and D had large variations, as shown in figures 5(c) and 5(d). In particular, the layer thickness of nacre D shows a large variation from 450 nm to 690 nm. Figure 6(a) shows the SEM images of the piece of nacre D, and figures 6(b) and 6(c) are magnified views in two parts of figure 6(a). The numbers in figures 6(b) and 6(c) are the number of aragonite layers. As is evident from these numbers in figures 6(b) and 6(c), we can easily confirm that the variation of the layer thickness is about 1.5 times. These SEM images show us how large the thickness variation of the nacre of Akoya oyster as a natural structure is.

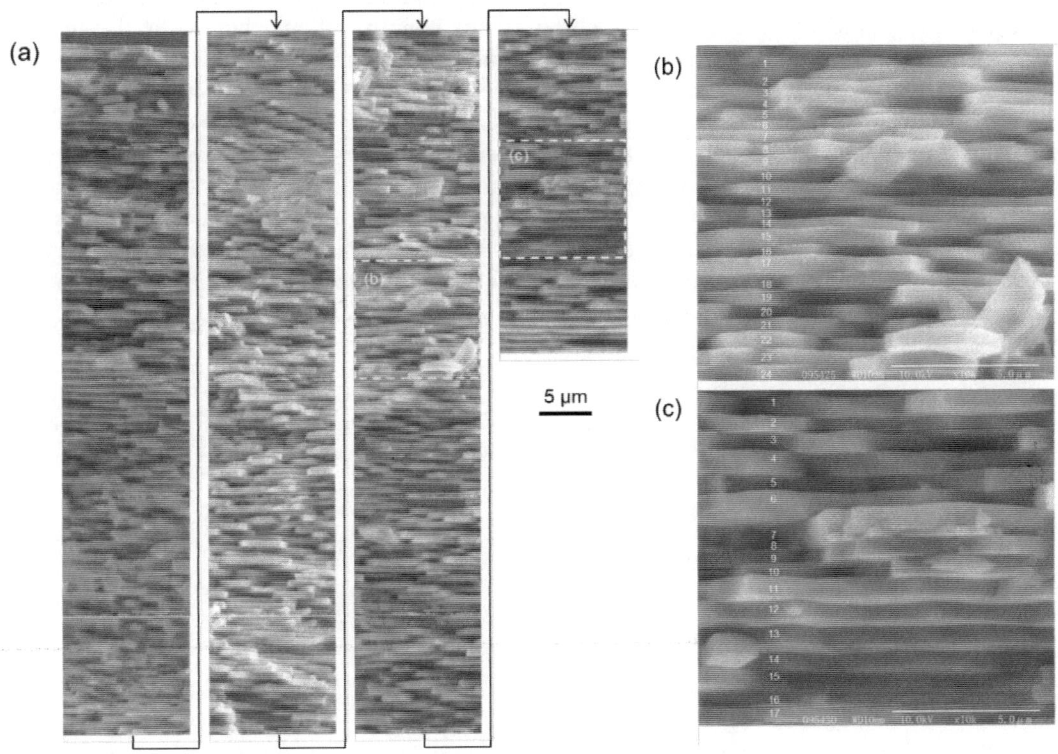

Figure 6. (a) SEM images of the piece of nacre D. (b) and (c) are magnified views.

4. Calculations of reflection spectra from Akoya pearl oyster

In this section, we again discuss the reflection properties of the pieces of nacre to connect the relation between the measured reflection spectra and the thickness profiles obtained by SEM observation. From the SEM images, we confirmed that the layer thicknesses of the nacre A and B were thinner than those of the nacre C and D. However, the reflection color of nacre A and B was approximately red, and that of nacre C and D was pale green. The pieces of nacre A and B having thinner layers showed us the longer wavelength reflection in the visible region. The reason is that the colors of the pieces of nacre A and B are caused by the second order reflection and that from the pieces of nacre C and D are caused by the third order reflection. The peak wavelength of the m-th order reflection λ_m is given by

$$\lambda_m = 2(n_1 d_1 + n_2 d_2)/m \tag{1}$$

where n_1 and d_1 are refractive index and the thickness of the aragonite crystal layer, and n_2 and d_2 are refractive index and the thickness of the conchiolin layer. When $d_1 = 400$ nm and $d_2 = 20$ nm, λ_2 and λ_3 are 691 and 461 nm, respectively. These wavelengths approximately match to the reflection spectra in figures 3(a) and 3(b) and the thickness profiles in figures 5(a) and 5(b). Here, we assumed n_1 and n_2 are 1.65 and 1.55. When $d_1 = 500$ nm and $d_2 = 20$ nm, λ_2 and λ_3 are 856, and 571 nm, respectively. These wavelengths also approximately match to the reflection spectrum in figure 3(c) and the thickness profile in figure 5(c).

The reflection spectra of the four pieces were calculated by transfer matrix method using thickness profiles obtained by SEM observation. The transfer matrix method is widely used for analysis of a one-dimensional layered structure. The data of layer thickness was obtained by spline interpolation. Figure 7 shows the measured and calculated reflection spectra for the four pieces of nacre. Red curves are calculated spectra which were filtered by Gaussian filter to reduce high order interference effect. In figure 7(a), although the calculated peak positions were slightly mismatched with the measured positions, the shape of the calculated curve showed a good agreement with that of experiment. Since the thicknesses of layers in the nacre are not uniform, depending on the *xy* position, we consider that the cause of the mismatch in peak position is the difference in measured area between optical and electron microscopes. On the other hand, the peak positions of the calculated reflection spectra in figures 7(b) and 7(c) showed a good agreement with those of experiment. The calculated curve in figure 7(d) did not show a clear large peak. This characteristic is similar to the measured spectrum. We consider that the reflection spectra calculated by transfer matrix using measured thickness profiles qualitatively agree with the experimental results. If light scattering could be simulated, the color of nacre of Akoya oyster would be reproduced in calculation. We believe such simulation is useful for pearl farming and grading of pearls.

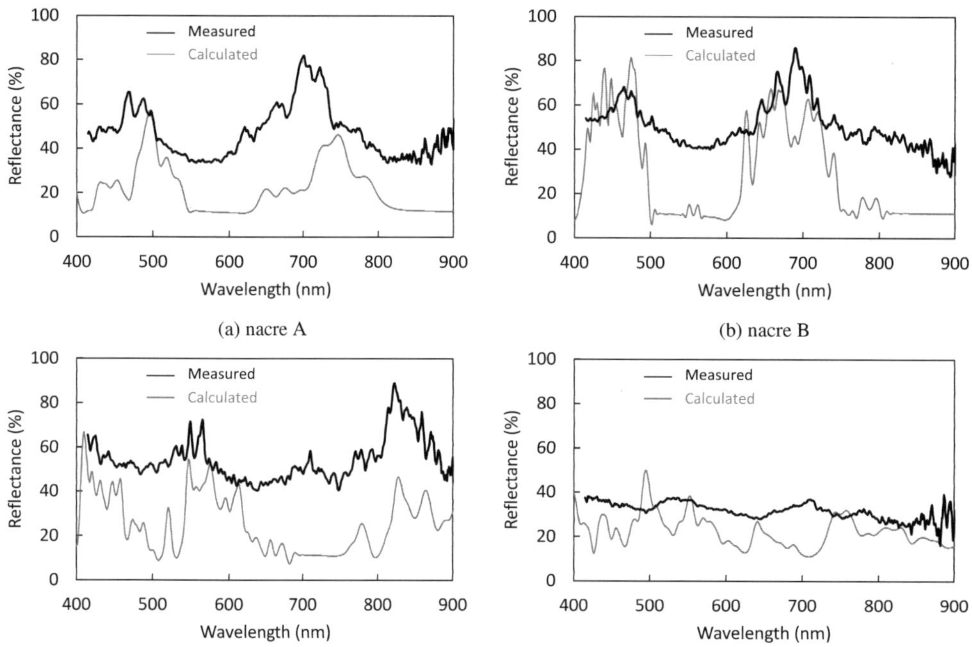

(a) nacre A (b) nacre B

(c) nacre C (d) nacre D

Figure 7. Measured and calculated reflection spectra of the pieces of nacre of Akoya oysters.

5. Conclusions

The reflection properties of the pieces of nacre of Akoya oysters were measured and the layer thickness profiles of the pieces were also observed by SEM. The layer thicknesses for red nacre were distributed at around 420 nm, and the thicknesses were approximately constant through the top to the bottom. In contrast, the thickness profiles of pale green nacre showed large variations. In particular, the layer thickness of nacre D showed a large variation from 450 nm to 690 nm. The reflection peaks in the measured spectra could be explained by Bragg theory and the results of SEM observation. Furthermore, the reflection spectra calculated by transfer matrix using measured thickness profiles qualitatively agreed with the experimental results. At the present stage, light scattering in nacre cannot be analyzed well. If light scattering could be simulated, the color of nacre of Akoya oyster would be reproduced in calculation. We believe that this study will contribute to the development of pearl farming and grading of pearls.

Acknowledgments
This research was supported by grants from the Project of the Bio-oriented Technology Research Advancement Institution, NARO (the special scheme project on advanced research and development for next-generation technology).

References

[1] Jiyu S, Bhushan B and Tong J 2013 *RSC Adv.* **3** 14862
[2] Yoshino K, Saitoh S, Shimoda Y, Kawagishi Y, Nakayama K and Ozaki M 1999 *Jpn. J. Appl.Phys.***38** L961
[3] Yoshino K, Shimoda Y, Kawagishi Y, Nakayama K and Ozaki M 1999 *App.Phys. Lett.* **75** 932
[4] Yablonovitch E 1987 *Phys. Rev. Lett.* **58** 2059
[5] John S 1987 *Phys. Rev. Lett.* **58** 2486
[6] Joannopoulos J D, Villeneuve P R and Fan S 1997 *Nature* **386** 143
[7] Dowling J P, Scalora M, Bloemer M J and Bowden C M 1994 *J. Appl.Phys.* **75** 944
[8] Painter O, Lee R K, Scherer A, Yariv A, O'Brien J D, Dapkus P D and Kim I 1999 *Science* **284** 1819
[9] Ozaki R, Matsui T, Ozaki M and Yoshino K 2003 *App. Phys. Lett.* **82** 3593
[10] Ozaki R, Shinpo T, Yoshino K, Ozaki M and Moritake H 2008 *App. Phys. Express* **1** 012003
[11] Foresi J S, Villeneuve P R, Ferrera J, Thoen E R, Steinmeyer G, Fan S, Joannopoulos J D, Kimerling L C, Smith H I and Ippen E P 1997 *Nature* **390** 143
[12] Noda S, Tomoda K, Yamamoto N and Chutinan A 2000 *Science* **289** 604
[13] Kinoshita S and Yoshioka S 2005 *ChemPhysChem* **6** 1442
[14] Odawara K, Ozaki R and Takagi M *J Fish Tech* in press (In Japanese)
[15] Iwanaga S, Yamada E, Kawaguchi T and Hosokawa H 2008 *Aquaculture Sci.* **56** 167 (In Japanese)

Negative capacitance in an organic solar cell observed by displacement current measurement

Koji Shimizu[1], Yuya Tanaka[1,2], Yutaka Noguchi[3] and Hisao Ishii[1,2,4]

[1]Graduate School of Advanced Integration Science, Chiba University, Chiba 263-8522, Japan
[2]Center for Frontier Science, Chiba University, Chiba, 263-8522, Japan
[3]School of Science & Technology, Meiji University, Kawasaki 214-8571, Japan
[4]Molecular Chirality Research Center, Chiba University, Chiba 263-8522, Japan

E-mail: y-tanaka@chiba-u.jp

Abstract. We applied displacement current measurement (DCM) to an organic solar cell (OSC) with a structure of ITO/CuPc/C_{60}/BCP/Al in order to investigate the origin of a negative capacitance (NC). In DCM curve, an anomalous characteristic that the current in backward scan was larger than that in forward scan was observed. We found that this characteristic originates from a gradual increase of an additional current, leading to NC in DCM. The additional current can be explained by the increase of the device temperature owing to a Joule heat and thermal deactivation of the exciton. Moreover, NC value was observed to be enhanced by light irradiation due to an additional temperature increase.

1. Introduction

Organic solar cells (OSCs) have been extensively studied toward practical use [1-3] since Tang invented a planar heterojunction OSC with a high power conversion efficiency (η) of 0.95% [4]. Advanced device structures such as p-type–intrinsic–n-type layers [5, 6], bulk-heterojunction [7, 8], and tandem cell structures [9, 10] have been developed to improve η. In addition, the insertion of additional layers between the electrode and organic layer has also enhanced the device efficiency and lifetime [11, 12]. It is progressing well, however, as the device structures gradually become complex, it has become difficult to clarify the operation mechanism, resulting in incomplete design guidelines for OSCs.

The investigation about the charge carriers in OSCs is valuable in order to understand the device physics because their behaviors, such as charge injection, dissociation, recombination, and extraction processes, determine the device performance. Thus the impedance spectroscopy (IS) [13, 14], time-of-flight (TOF) [15, 16], charge extraction by linearly increasing voltage (CELIV) [15, 17], and displacement current measurements (DCM) [18] have been applied to evaluate them. Among these techniques, IS is the most widely used one and the device parameters have been analyzed based on, mainly, the frequency dependence of the capacitance of the device.

In IS measurements of organic devices, a negative capacitance (NC) is often observed at high bias and low frequency, and has been explained by various models, e.g., the recombination current [19], electron injection through interfacial states [20, 21], charge trapping [22], bias dependent conductivity modulation [23], etc. Recently, a self-heating model has been proposed by Okumoto and Tsutsui [24], and Knapp and Ruhstaller have confirmed the validity by simulation [25]. In addition, they found that NC originates from self-heating induced current enhancement. In IS, the transient response cannot be correctly evaluated because it does not show a periodical change [26]. Although the advancement in the comprehension of NC is made, no studies have ever tried to evaluate the NC phenomenon in OSCs based on the self-heating model.

In this study, we applied DCM to a prototypical OSC with a structure of ITO/CuPc/C_{60}/BCP/Al, in order to investigate the origin of NC through the direct measurement of the flowing current. We found that NC observed in DCM originates from an anomalous characteristic that the current in the backward (positive to negative bias) scan was higher than that in the forward (negative to positive bias) scan. A transient current measurement revealed that this inversion, which does not normally occur, arises from a gradual increase in current owing to the self-heating effect of the device, leading to the NC in DCM. Finally, an impact of the light irradiation on NC is discussed based on the properties of the transient current.

2. Experimental

We prepared a OSC with a structure of ITO (50 nm)/ copper phthalocyanine (CuPc) (30 nm)/ C_{60} (40 nm)/ bathocuproine (BCP) (10 nm)/ Al (100 nm). Depositions of the organic materials and metal were done in the dark at the deposition rate of typically 1 Å/s and 3-5 Å/s, respectively, with a pressure of below 10^{-4} Pa. An active area of organic layer was 4 mm^2. After the deposition, the device was transferred to the glove box filled with pure N_2, and encapsulated without exposure to air.

In DCM, a triangular wave voltage (V) is applied to the sample, and the current response which is the sum of the actual current (I_{act}) and the displacement current (I_{dis}) is measured. Because I_{dis} is proportional to the effective capacitance (C) of the device as well as a ramp-rate (dV/dt) in quasi-static state ($I_{dis} = C(dV/dt)$), the total current in forward scan (I^+) can be expressed by

$$I^+ = I_{act} + C\left|\frac{dV}{dt}\right|. \qquad (1)$$

Similarly, the total current in backward scan (I^-) is given by

$$I^- = I_{act} - C\left|\frac{dV}{dt}\right|. \qquad (2)$$

From the difference between the equations (1) and (2), C can be obtained as

$$C = \frac{(I^+ - I^-)}{2|dV/dt|}. \qquad (3)$$

C normally shows a positive value since I^+ is larger than I^- because of $C|dV/dt|$ term.

The electrical measurements were performed in the dark or under illumination with an AM 1.5 solar simulator (Asahi spectra, HAL-320) at room temperature. Note here that a cold filter was inserted between the device and a solar simulator in order to suppress the increase of the device temperature by an infrared light. For the DCM measurements, the measurement system for ferroelectrics (Toyo Corporation, FCE-1) was employed. The voltage sweep was started from the minimum voltage of the triangular wave (V_{min}) in all DCM experiments.

3. Results and discussion

3.1. DCM evaluation for organic solar cell

Figure 1 shows DCM curves in the dark at the ramp-rate of 0.1 V/s. V_{min} and the maximum voltage (V_{max}) of the triangular wave were set to be –0.5 V and 0.5 V, respectively. Solid and dotted lines indicate I^+ and I^-, respectively. The expanded figures of the regions (I), (II), and (III) are shown in the insets, clearly indicating the difference between I^+ and I^-. In region (I), C was a positive value because of $I^+ > I^-$. Intriguingly, I^+ and I^- intersected at 445 mV in region (II) and finally I^- became larger than I^+ in region (III). These results imply that C which is obtained from the equation (3) varied from a positive to a negative value at 445 mV.

IOP Conf. Series: Journal of Physics: Conf. Series **924** (2017) 012012 doi:10.1088/1742-6596/924/1/012012

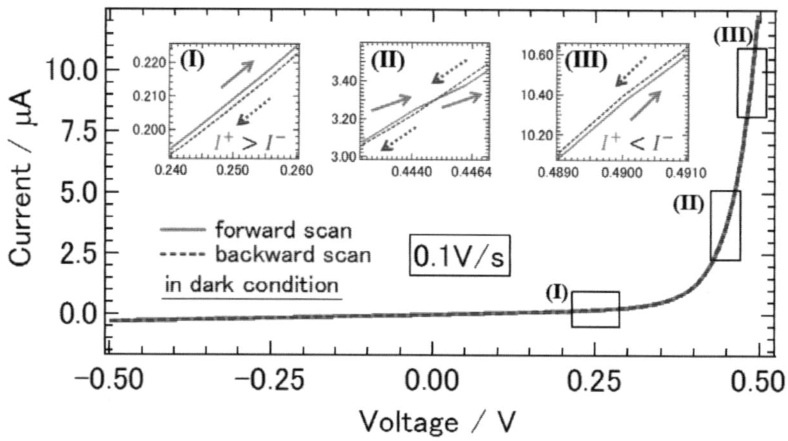

Figure 1. DCM curves in forward (solid line) and backward (dotted line) scans at the ramp-rate of 0.1 V/s. The insets indicate the expanded figures of the regions (I), (II), and (III).

Figure 2(a) indicates V dependence of C calculated from the equation (3) (CV curve), and V_{max} was changed from 0.5 V to 1.5 V in 0.2 V step. The inset shows the expanded figure of DCM curve at V_{max} of 0.5 V, showing that C became negative at higher voltage than 445 mV. In these way, NC phenomenon of OSC was observed in DCM as well as IS measurements. In figure 2(a), at first, C decreased with increasing V from 445 mV and reached the minimum value of C (C_{min}). Then C increased with V and finally became zero at V_{max}. In addition, the absolute value of C_{min} ($|C_{min}|$) increased with increasing V_{max}. In order to directly evaluate V_{max} dependence in more detail, I^+ and I^- were separately plotted as shown in the left and right figures of the figure 2(b), respectively. I^+ coincided well each other, while I^- increased with increasing V_{max}. All these things make it clear that V_{max} dependence of C_{min} originates from the increase of I^-, indeed, an additional current (I_{add}) which prominently appears during the backward scan and depends on V_{max}.

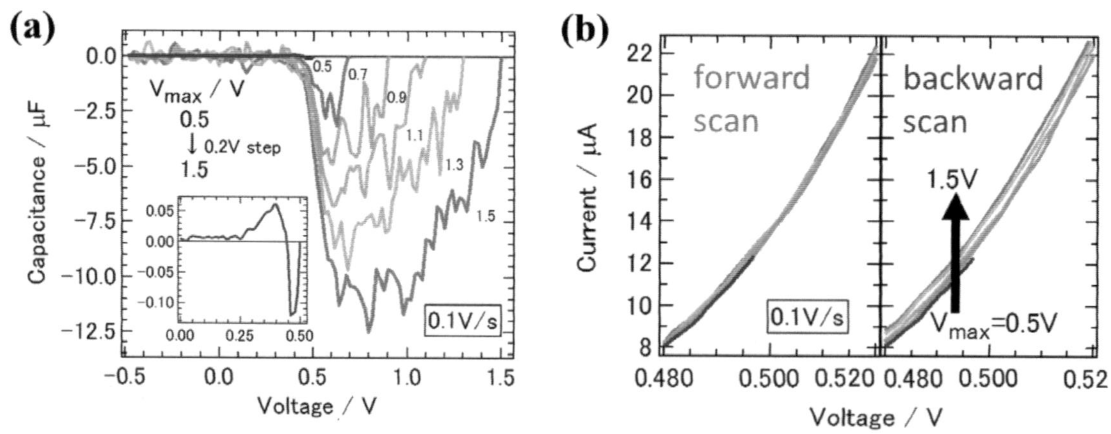

Figure 2. (a) CV curves when V_{max} was changed from 0.5 V to 1.5 V in 0.2 V step. The inset indicates the CV curves at V_{max} of 0.5 V. (b) V_{max}-dependence of DCM curves in forward (left figure) and backward (right figure) scans.

One explanation for the origin of I_{add} may be a transient response of the current. Figure 3(a) shows the ramp-rate dependence of C at V_{max} = 1.5 V, and the ramp-rate was varied from 0.1 V/s to 4 V/s. C_{min} drastically increased with an increase in the ramp-rate, suggesting that I_{add} strongly depends on the time.

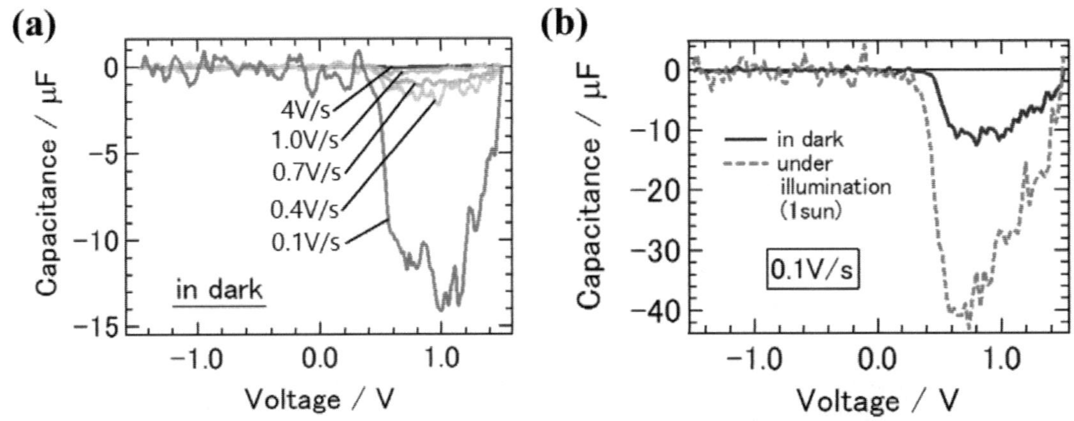

Figure 3. (a) Ramp-rate dependence of CV curves in the dark. (b) CV curves in the dark (solid line) and under illumination (dotted line).

Before turning to the discussion about the origin of NC, we must draw attention to the effect of light irradiation. Figure 3(b) shows CV curves in the dark (solid line) and under illumination (dotted line). The shape of CV curve under illumination was similar to that in the dark, while $|C_{min}|$ drastically increased by light irradiation. Thus, it is clear that I_{add} is strongly affected by the light irradiation.

Finally, the features of I_{add} mentioned above can be summarized as follows:

(i) I_{add} prominently appears during the backward scan (figure 1),
(ii) I_{add} increases with increasing V_{max} (figures 2(a) and 2(b)),
(iii) I_{add} decreases with increasing the ramp-rate (figure 3(a)),
(iv) I_{add} is enhanced by light irradiation (figure 3(b)).

The discussion about each feature will be presented in the next section.

3.2. Origin of negative capacitance value

As discussed in the previous section, it is likely that NC originates from the increase of I^-, namely, I_{add} over time. The details of the time dependence of the current can be easily evaluated by keeping the voltage at constant value (V_k) [26]. Solid line in the figure 4 indicates the time dependence of the current in the dark when the rectangular wave voltage was applied (at 0 s, the voltage was changed to $V_k = 1$ V and held as shown in the inset). The vertical axis is the normalized value (I_n) by the initial current ($t = 0$ s). In the dark, I_n increased by I_{add} with t. Taking this into consideration, DCM and CV curves can be explained as follows: Firstly, I_{add} starts to flow from a certain voltage (V_s) in forward scan (figure 1), and gradually increases with t. Because I_{add} continues to increase even in backward scan if the applied voltages are higher than V_s, I^- becomes larger than I^+, resulting in the appearance of NC (feature (i)). Since I_n was not saturated within 100 s (figure 4), in the case of the same ramp-rate, the higher V_{max} is, the larger I_{add} flows (figure 2(a)), leading to the enhancement of $|C_{min}|$ with V_{max} (feature (ii)). In the same way, C_{min} increased with increasing the ramp-rate (figure 3(a)) because I_{add} becomes smaller by shortening of the time (feature (iii)). One can be fairly certain that the NC observed in DCM originates from I_{add} which depends not directly on the applied voltage and current at which NC appears, but on the time and V_{max}.

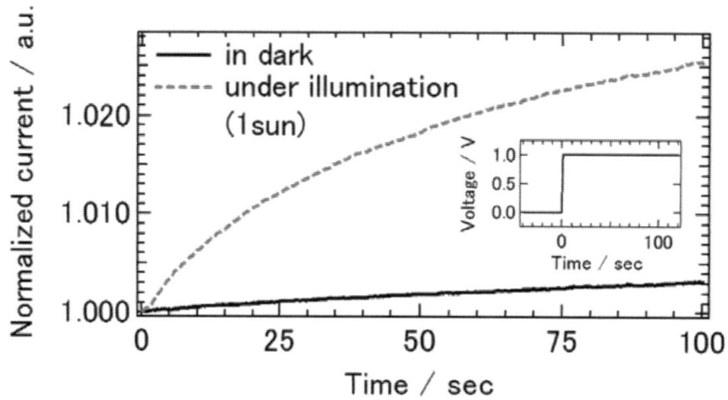

Figure 4. Transient current measurements in the dark (solid line) and under illumination (dotted line). The inset indicates the waveform of applied voltage which is kept at 1 V after 0 s. The light illumination was started a few seconds before V_k became 1 V.

In order to discuss the origin of I_{add}, it is important to consider the transport mechanism of charge carriers because the appearance of I_{add} actually implies the variation of conductance of the device. In organic semiconductors, charges spend most of their time on localized deep states where they cannot contribute to current [27]. It must be noted that the trap states are not steadily filled with the charges, indeed, the states repeatedly trap and release the charges in the transport level [21,28]. One example of this is Poole-Frenkel conduction, which includes temperature- and field-dependent mobility [28]. Because I_{add} increases even at the constant applied voltage as shown in the figure 4, it is likely that I_{add} was not enhanced by the applied field, in other words, it is clear that the conductance was changed by the temperature variation of the device.

The most likely explanation for the temperature variation of the device is that the heat is generated by the flowing current (Joule heat) and a thermal deactivation of the exciton, that is, a self-heating occurs. In the framework of Poole-Frenkel conduction, the current increases with temperature because the charges can be thermally excited from the trap states into the transport level. Since the I_n was not saturated within 100 s, I^- becomes higher than I^+ immediately after the start of the backward scan (figure 1), resulting in the appearance of I_{add} (feature (i)). In like manner, I_{add} increases with V_{max} (figure 2(a)) because the device temperature continues to increase (feature (ii)). On contrary, I_{add} decreases with increasing the ramp-rate (figure 3(a)) since there is not enough time to increase the device temperature (feature (iii)).

We are now ready to consider the effect of the light irradiation on NC value observed in DCM (feature (iv)). Although the shape of CV curve under illumination was approximately in agreement with that in the dark, $|C_{min}|$ drastically increases by light irradiation as shown in the figure 3(b). Dotted line in the figure 4 indicates the time dependence of the current under illumination when the rectangular wave voltage (V_k = 1 V) was applied at 0 s. The light irradiation was started a few seconds before V_k became 1 V. This result shows that I_{add} under illumination was much larger than that in the dark. It seems reasonable to suppose that the enhancement of I_{add} by light irradiation leads to the increase of $|C_{min}|$ under illumination.

As discussed above, again, the two causes of the conductance variation are considered to be the temperature and the applied field. One can safely state that no field-assisted variation occurs because the applied voltage was fixed at 1 V. These results imply that the device temperature is increased by light irradiation, resulting in the enhancement of NC value.

4. Conclusions

In DCM, we investigated the origin of NC phenomenon of OSCs with a structure of ITO/ CuPc/ C_{60}/ BCP/ Al. We found an anomalous characteristic that the current in backward scan is larger than that in forward scan, leading to the appearance of NC in DCM. Transient current measurement revealed that this characteristic originates from the gradual increase of the additional current with time. This current can be explained by the increase of device temperature resulting from a self-heating effect. Moreover, the absolute value of NC observed in DCM drastically increased by the light irradiation. From the comparison of transient current in the dark and under illumination, we concluded that the device temperature was increased by the light irradiation, leading to the enhancement of NC value. Because of the low power conversion efficiency of the OSCs, the major part of the input power by light irradiation would be changed to the heat due to the thermal deactivation of the excitons. Thus we believe that the detailed analysis of NC phenomenon by DCM leads to the understanding of the charge recombination mechanism in OSCs.

Acknowledgements

This work was supported by the Global-COE program at Chiba University (Advanced School for Organic Electronics, G-03, MEXT), JSPS KAKENHI (Grants Nos. 21245042, 16H04222, 16H06670), and the Funding Program for World-Leading Innovative R&D on Science and Technology (FIRST). YT gratefully acknowledges JSPS Research Fellowships for Young Scientists (Grant No. 22·7042).

References

[1] Peumans P, Yakimov A and Forrest S R 2003 *J. Appl. Phys.* **93** 3693–723
[2] Hains A W, Liang Z, Woodhouse M A, and Gregg B A 2010 *Chem. Rev.* **110** 6689–735
[3] Jørgensen M, Norrman K, Gevorgyan S A, Tromholt T, Andreasen B and Krebs F C 2012 *Adv. Mater.* **24** 580–612
[4] Tang C W 1986 *Appl. Phys. Lett.* **48** 183–5
[5] Hiramoto M, Fukusumi H and Yokoyama M 1991 *Appl. Phys. Lett.* **58** 1062–4
[6] Xue J, Rand B P, Uchida S, and Forrest S R 2005 *Adv. Mater.* **17** 66–71
[7] Yu G, Gao J, Hummelen J C, Wudl F and Heeger A J 1995 *Science* **270** 1789–91
[8] Halls J J M, Walsh C A, Greenham N C, Marseglla E A, Friend R H, Moratti S C and Holmes A B 1995 *Nature* **376** 498–500
[9] Xue J, Uchida S, Rand B P and Forrest S R 2004 *Appl. Phys. Lett.* **85** 5757–9
[10] Yu B, Zhu F, Wang H, Li G and Yan D 2008 *J. Appl. Phys.* **104** 114503
[11] Peumans P, Bulovic V and Forrest S R 2000 *Appl. Phys. Lett.* **76** 2650–2
[12] Peumans P and Forrest S R 2001 *Appl. Phys. Lett.* **79** 126–8
[13] Glatthaar M, Riede M, Keegan N, Sylvester-Hvid K, Zimmermann B, Niggemann M, Hinsch A and Gombert A 2007 *Sol. Energ. Mat. Sol. C.* **91** 390–3
[14] Garcia-Belmonte G, Boix P P, Bisquert J, Sessolo M and Bolink H J 2010 *Sol. Energ. Mat. Sol. C.* **94** 366–75
[15] Pivrikas A, Sariciftci N S, Juška G and Österbacka R 2007 *Prog. Photovolt: Res. Appl.* **15** 677–96
[16] Tanaka Y, Noguchi Y, Oda K, Nakayama Y, Takahashi J, Tokairin H and Ishii H 2014 *J. Appl. Phys.* **116** 114503
[17] Juška G, Arlauskas K, Österbacka R and Stubb H 2000 *Synth. Met.* **109** 173–6
[18] Tanaka Y 2013 *Ph.D. thesis Chiba University*
[19] Ehrenfreund E, Lungenschmied C, Dennler G, Neugebauer H, and Sariciftci N S 2007 *Appl. Phys. Lett.* **91** 012112
[20] Bisquert J, Garcia-Belmonte G, Ángeles P and Bolink H J 2006 *Chem Phys. Lett.* **422** 184–91
[21] Bisquert J 2008 *Phys. Chem. Chem. Phys.* **10** 3175–94
[22] Kwok H L 2003 *Solid-State Electronics* **47** 1089–93
[23] Pingree L S C, Scott B J, Russell M T, Marks T J and Hersam M C 2005 *Appl. Phys. Lett.* **86** 073509
[24] Okumoto H and Tsutsui T 2014 *Appl. Phys. Express* **7** 061601
[25] Knapp E and Ruhstaller B 2015 *J. Appl. Phys.* **117** 135501

[26] Tanaka Y, Noguchi Y, Kraus M, Brütting W, Ishii H 2011 *Org. Electron.* **12** 1560–5
[27] Stallinga P 2009 *Electrical Characterization of Organic Electronic Materials and Devices* (Padstow: Wiley) chapter 1 pp 6–10
[28] Stallinga P 2011 *Adv. Mater.* **23** 3356–62

Effect of immobilization technique on performance ZnO nanorods based enzymatic electrochemical glucose biosensor

Mayoorika Shukla[a], Pramila[a], I.A. Palani[b,c] and Vipul Singh[a,c*]
[a] Molecular and Nanoelectronics Research Group (MNRG), Discipline of Electrical Engineering, Indian Institute of Technology Indore, Indore-453552, Madhya Pradesh, India.
[b] Mechatronics and Instrumentation Lab, Discipline of Mechanical Engineering, Indian Institute of Technology Indore, Indore-453552, Madhya Pradesh, India.
[c] Centre for Material Science and Engineering, Indian Institute of Technology Indore, Indore-453552, Madhya Pradesh, India.
E-mail: *vipul@iiti.ac.in

Abstract: In this paper, ZnO Nanorods (ZNR) have been synthesized over Platinum (Pt) coated glass substrate with in-situ addition $KMnO_4$ during hydrothermal growth process. Significant variation in ZnO nanostructures was observed by $KMnO_4$ addition during the growth. Glucose oxidase was later immobilized over ZNRs. The as-prepared ZNRs were further utilized for glucose detection by employing amperometric electrochemical transduction method. In order to optimize the performance of the prepared biosensor two different immobilization techniques i.e. physical adsorption and cross linking have been employed and compared. Further investigations suggest that immobilization via cross linking method resulted in the improvement of the biosensor performance, thereby significantly affecting the sensitivity and linear range of the fabricated biosensor. Among the two types of biosensors fabricated using ZNR, the best performance was shown by cross linked electrodes. The sensitivity for the same was found to be 17.7 $mA\text{-}cm^{-2}\text{-}M^{-1}$, along with a wide linear range of 0.5-8.5 mM.

Keywords – Physical Adsorption, Cross linking, ZnO Nanorods, Glucose biosensor.

1. Introduction

Biosensors have huge applications in wide range of fields from health care, environmental monitoring to industries. It can be defined as a device which perceives bio-recognition event for detection of analyte carried out by biomolecules immobilized over support matrix. The output signal of such events is proportional to the analyte concentration [1]. Among the different categories of biosensors, enzymatic glucose biosensor has attracted interest of researchers due to its importance in diabetes care, beverage and food industry, drug discovery and analysis [2]. The body cells utilize glucose both as a source of energy and as a metabolic intermediate in the synthesis of more complex molecules such as fat [3]. Diseases like diabetes mellitus are result of imbalance in blood glucose levels. Owing to poor lifestyle and the resulting obesity thereby, has led to tremendous increase in the number of patients suffering from diabetes in the last few decades which has led to high demand for reliable point of care testing devices. In this regard, nanostructures of semiconductor metal oxides such as ZnO, SnO_2, V_2O_5, and TiO_2 etc. have been extensively explored in development of biosensors due to their multifunctional properties [4, 5]. Morphology of these materials in reduced dimensions has strong influence over its aspect ratio and electrical characteristics which in turn are beneficial for its usage as a support matrix. Amid all these, ZnO has been proven to be a good candidate as support matrix for immobilization of enzymes because of its low toxicity, biocompatibility, fair thermal and oxidation stability, high electron mobility, high iso-electric point (IEP=9.5) and amicable tunability into wide variety of nanostructures [6-8]. ZnO is a wide band gap (E_g 3.3 eV at 300K) II-VI compound semiconductor with high exciton binding energy of 60 meV at room temperature [9]. ZnO can be grown in a variety of morphologies viz. nanorods, nanowires, nanobelts, nanosheets, nanotubes, etc. by different physical (chemical vapor deposition, pulsed laser

IOP Conf. Series: Journal of Physics: Conf. Series **924** (2017) 012013 doi:10.1088/1742-6596/924/1/012013

deposition, spray pyrolysis etc.) and solution based (hydrothermal, sol-gel, electrochemical etc.) fabrication techniques [10-12].

Biosensors utilizing ZnO as support matrix for enzyme immobilization have been explored exhaustively in literature for detection of glucose, urea, cancer cells, cholesterol etc. [13-17]. Enzyme immobilization refers to confinement enzymes to the support matrix with retained catalytic activity. The major components of enzyme immobilization are enzyme, support matrix and mode of attachment. The mode of attachment can be broadly classified into four categories i.e. physical adsorption, covalent Binding, co-entrapment and cross linking each having its own merits [18]. The choice of immobilization technique and nanostructure plays key role in performance of biosensors as it directly affects enzyme loading which eventually is responsible for improvement in biosensors performance. ZnO thin films and a wide variety of ZnO nanostructures have been used for glucose biosensing applications. However owing to low aspect ratio of ZnO thin films, nanostructures such as nanorods, nanocombs, nanotubes etc. are preferred over the same for improved performance of biosensor. As high aspect ratio leads to higher enzyme loading which holds key to improved biosensor performance. In literature rf-sputtered ZnO thin films have been reported where sensitivity was obtained around 0.5-1.3 $mA\text{-}cm^{-2}\text{-}M^{-1}$ [19]. Unlike this, ZnO nanostructures have been found to have improved sensitivities around 12-15 $mA\text{-}cm^{-2}\text{-}M^{-1}$ [20].

In this work, an electrochemical enzymatic glucose biosensor has been analyzed employing two different types of immobilization methods *viz.* physical adsorption and cross-linking. Glucose oxidase (GO_x) was used for amperometric detection of glucose. ZNRs synthesized by hydrothermal growth technique were utilized as a support matrix for immobilization of GO_x. It is a cost effective and low temperature process which enables easy large scale synthesis of ZnO nanostructures. The high IEP feature of ZnO enables physical adsorption of GO_x over it by electrostatic binding while cross linking has been performed by using glutaraldehyde and bovine serum albumin as a cross linking agent and stabilizer respectively [13, 21]. The two techniques were compared here towards optimization of biosensor performance.

2. Experimental Section

2.1 Materials

Glucose oxidase from Aspergillus niger (E.C. 1.1.3.4), D-(+)-glucose, Glutaraldehyde, Bovine serum albumin, Nafion (5 wt % in water and lower aliphatic alcohol), anhydrous Zinc Acetate, Ethanolamine, 2-methoxy ethanol, Zinc Nitrate Hexahydrate ($Zn(NO_3)_2.6H_2O$), Hexamethylenetetramine ($C_6H_{12}N_4$), $KMnO_4$, Sodium phosphate dibasic (Na_2HPO_4), Potassium phosphate monobasic (KH_2PO_4), NaCl, Potassium ferricyanide/ferrocyanide $[Fe(CN)_6]^{-3/-4}$ were used as procured from Sigma Aldrich. All chemicals were of analytical grade and used without any further purification. 0.01M Phosphate buffered saline solution was prepared from Na_2HPO_4 and KH_2PO_4 with 0.137 mM NaCl, and pH was adjusted to 7.4. Glucose stock solution was kept overnight for mutarotation after preparation, in order to establish equilibrium concentration between α and β anomers. De-ionized (DI) water (>18 MΩ) obtained from Milli-Q purifying system was used in preparation of all the solutions.

2.2 Synthesis of ZNR, enzyme immobilization and quantification

Glass samples cut into 1 cm x 1 cm size were cleaned by ultrasonication process in acetone, isopropyl alcohol and DI respectively followed by drying. After cleaning Platinum (Pt) was sputtered with a 5 nm underlying buffer layer of Cr, at base pressure of $1x10^{-6}$ mbar. Cr buffer layer was incorporated to improve adhesion of Pt over glass [22]. Samples were rotated at a speed of 10 revolutions per minute (rpm) to ensure uniform deposition of Pt over it. During the growth process, initially seed layer was coated over Glass/Cr:Pt samples followed by hydrothermal growth of ZNRs as reported in our previous work [23]. In summary, an equimolar solution of Zinc acetate and ethanolamine in 2-methoxy ethanol has been spun coated followed by annealing for seed layer preparation. Then, seed layer deposited samples

IOP Conf. Series: Journal of Physics: Conf. Series **924** (2017) 012013 doi:10.1088/1742-6596/924/1/012013

were kept for growth of ZNR in 0.1M equimolar solution of Zinc nitrate and Hexamethylenetetramine (HMTA) with 5 mM $KMnO_4$ as additive. On completion of growth, samples were cleaned thoroughly with DI followed by drying to remove any residual salt present.

Finally after growth of ZNR two types of electrodes were prepared using physical adsorption and cross linking immobilization method. In physical adsorption, 5µl solution of 20 mg/ml GO_x prepared in PBS was dropped over ZNR and kept in incubator for 24 h. After this GO_x immobilized ZNR was washed with PBS to remove unadhered enzymes from the electrode. Further, 0.5% Nafion solution was drop casted over it and kept overnight for drying. However for employing cross linking process, 100 µL GO_x solution, 50 µL (2.5 %) glutaraldehyde and 50 µL (0.5 %) Nafion solution was mixed thoroughly. After this 4 µL of the as prepared solution was drop casted over electrode and allowed to dry at room temperature. Finally, 2 µL 0.5 % Nafion solution was coated to avoid leaching of enzymes. After drying of nafion solution electrodes were cleaned for removal of unadhered GO_x molecules. Both the type electrodes were kept in incubator at 4 °C when not in use [21].

An aromatic residue namely tryptophan is present in each GO_x molecule which contributes to an intrinsic fluorescence in it. Therefore GO_x solution when excited at 280 nm gives fluorescence emission at 330 nm. Emission intensity at 330 nm is proportional to the amount of GO_x present in a solution [2]. This property of GO_x has been well utilized for determining amount of immobilized enzymes over support matrix i.e. enzyme loading. The linear calibration curve for known amount of GO_x with its corresponding fluorescence intensity was later obtained. PBS in which each type of electrode has been cleaned after immobilization and solution was collected separately for taking fluorescence. A known amount of GO_x i.e. 100 µg has been immobilized over support matrix. Thus amount of unadhered GO_x can be determined by fluorescence intensity of collected solution and calibration curve. Finally an estimate of the GO_x loaded over support matrix can be made from the difference in the values of the two.

2.3 Characterization and measurement

Morphology and structures of ZnO dumbbells and ZNR were studied by FESEM (Sigma Supra^{TM} 55; Carl Zeiss). The crystalline structure and phase transformations were analyzed by powder XRD (RigakuSmart Lab® System) with Cu Kα radiation operating at 40 kV and 40 mA, using Cu Kα radiation (λ = 1.5418 Å) in 2θ range from 20° to 80°. The PL spectrometer (Dongwoo Optron DM 500i) having an excitation source consisting of a continuous wave He-Cd laser (excitation wavelength, 325 nm, PMT detector) was used to measure the PL emission from these samples. Fluorescence emissions were obtained using FluoroMax-4p Spectrofluorometer from Horiba Jobin Yvon (Model: FM-100). Excitation and emission slit widths were set at 2 nm each and the samples were excited at 280 nm. Image processing toolbox of MATLAB has been used to study length and diameter distribution of as prepared ZNR. Electrochemical measurements were performed by electrochemical work station (Autolab PGSTAT 302N) using three electrode cell configuration with Ag/AgCl and Pt foil electrode as reference and counter electrode respectively.

3. Results and discussion

3.1 Surface Morphology and Structural Characterization –

The morphology of obtained ZnO nanostructures were analyzed by FESEM which is a non-contact mode technique generally used for surface characterization of films. Figure 1(a) shows the image of ZnO microstructures which were grown in pristine solution i.e. without addition of $KMnO_4$. The microstructures in form of dumbbells were observed with the growth in all the directions. Apart from this grain boundary was observed to be in the middle of microstructures which were ranging from 3 µm to 6 µm [25]. Figure 1 (b-c) shows the top view and cross sectional view of ZNR grown with the addition of 5 mM $KMnO_4$. The as prepared nanorods were found to have average diameter and length distribution to be around 130 nm and 530 nm, respectively. The nanorods were uniformly distributed over the entire substrate. The mechanism of growth of ZNR with $KMnO_4$ addition has been discussed in detail in our previous report [23]. Briefly, presence of additive has strongly influenced hydrolysis of HMTA and ceased formation of dumbbells, while the oxygen rich environment in the nutrient solution has led to the passivation of the defects.

Figure 1. FESEM images of (a) ZnO dumbbells, (b) top view and (c) cross sectional view of ZNR.

The XRD pattern for as prepared ZnO dumbbells and ZNR grown with 5 mM $KMnO_4$ is shown in Fig. 2. Both ZnO dumbbells and ZNR were found to consist of all the diffraction peaks corresponding to ZnO wurtzite phase (space group P63mc, JCPDS – 36 - 1451). An extra peak indexed to Pt coated over glass substrate at $2\theta = 40°$ indexed. The c-axis (002) peak intensity was observed to enhance on addition of $KMnO_4$ since presence of strong oxidizing agent acts as a driving force for anisotropic growth which provides an oxygen rich environment. There was no impurity or intermediate formation during growth in the sample as no peak other than ZnO was present within the detection limit of the instrument. The as prepared dumbbells and ZNR were well crystalline owing to sharp and intense diffraction peaks.

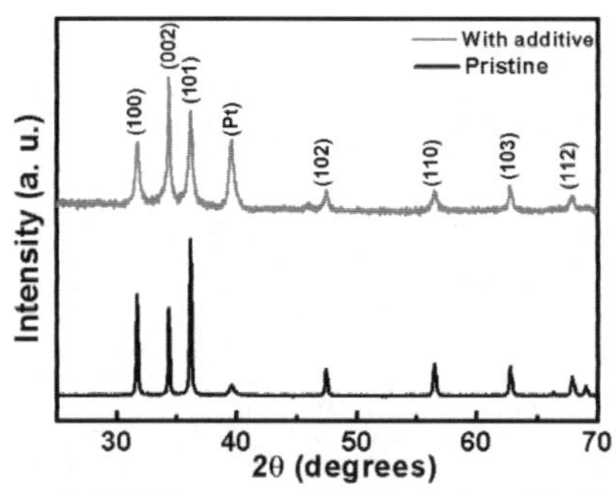

Figure 2. XRD pattern for ZnO dumbbells and ZNR.

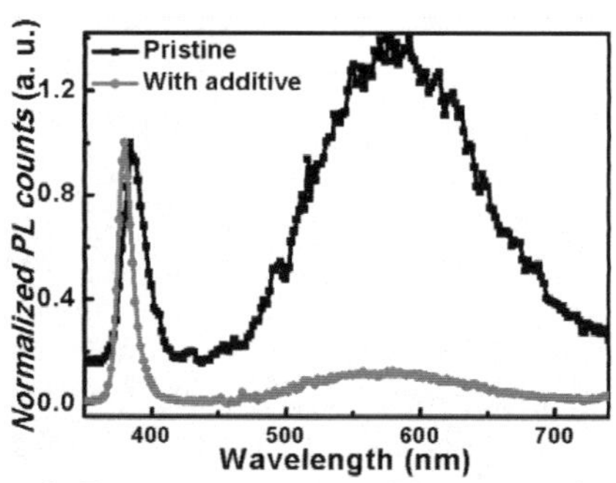

Figure 3. Room temperature PL spectra for ZnO dumbbells and ZNR.

3.2 Optical Characterization

The optical characterization of ZnO dumbbells and ZNR was performed by the room temperature Photoluminescence (PL) measurements. Generally PL of ZnO consists of two peaks i.e. Near band edge emission (NBE) in ultra violet region and defect level emission (DLE) in visible region. The NBE corresponds to excitonic transitions around band gap of ZnO. The broad peak associated with DLE originates by presence of more than one type of defects [26, 27]. The actual reason behind DLE is still not fully understood. The normalized PL spectra for the two types of nanostructures obtained with and without addition of $KMnO_4$ have been shown in Fig. 3. It consists of two peaks; a sharp NBE around 385 nm and a broad DLE ranging from 450 nm to 750 nm centered around 580 nm. It is interesting to note that DLE got suppressed by incorporation of $KMnO_4$ during the growth which creates an oxygen rich environment [23]. Besides this a blue shift in UV peak was observed in samples grown with additive due to Burstein-Moss effect. In n-type semiconductors, when carrier concentration reaches beyond a limit, Fermi level moves into the conduction band due to filling of conduction band by electrons [28-30]. PL study has revealed a clear repression of defects by $KMnO_4$ addition.

3.3 Enzyme quantification

Enzyme loading was analyzed by optical characteristic of GO_x molecules due to the presence of tryptophan which is responsible for intrinsic fluorescence in it. Aspect ratio has been calculated as ratio of average length to average diameter and was found to be 4.2 for as prepared ZNR. Enzyme loading depends on the surface to volume ratio available for immobilization of biomolecules and mode of immobilization. Enzyme loading was measured to be 425 $\mu g\text{-}cm^{-2}$ and 575 $\mu g\text{-}cm^{-2}$ for physical adsorption and cross linking immobilization method respectively. Physical adsorption utilizes electrostatic attraction between GO_x molecules (IEP=4.5) and ZnO (IEP=9.5) at neutral pH value while in case of cross linking, enzymes join in 3D structure by covalent bond formation between enzyme (GO_x) and bi-functional cross linking agent (Glutaraldehyde) [2]. Thus, cross linking has resulted in higher enzyme loading as compared to physical adsorption.

3.4 Biosensor Characterization

3.4.1 Cyclic Voltammetry Studies

All the electrochemical experiments were performed using three electrode cell configuration. Glass/Cr:Pt/ZNR/GOx/Nafion, Ag/AgCl and Pt foil were utilized as working, reference and counter electrode respectively. PBS with 0.01 M concentration and pH=7.4 was used as electrolyte solution. The electrochemical characterization of different biosensors was investigated by Cyclic Voltammograms (CV) to study catalytic oxidation of glucose. CV was recorded in an unstirred PBS solution at scan rate of 100 mV/s swept from -0.2V to 0.8V. Figure 4 (a) shows CV for the cross linked ZNR electrodes in absence and on subsequent addition of glucose. As compared to the background in absence of glucose a sharp increase in the anodic current for cross linked electrode in presence of glucose has been observed. It was clearly evident from the CV that anodic oxidation current has increased significantly and cathodic reduction current was suppressed indicating catalytic oxidation of glucose. For cross linked GO_x electrodes equal amount of glucose was injected from 3 mM to 12 mM with an increment of 3 mM. It was noticeable from it that almost same increment in anodic current was observed till 9 mM glucose injection which suggests that the biosensor exhibits extended linear range of operation. Further CV was obtained for the two types of electrodes in presence of 3 mM glucose in PBS as shown in Fig. 4 (b). The anodic current was significantly enhanced in case of cross linked electrodes as compared to physically adsorbed electrodes owing to the fact that enzyme loading has increased. Further to explore findings obtained by CV, amperometric biosensor response was recorded.

IOP Conf. Series: Journal of Physics: Conf. Series **924** (2017) 012013 doi:10.1088/1742-6596/924/1/012013

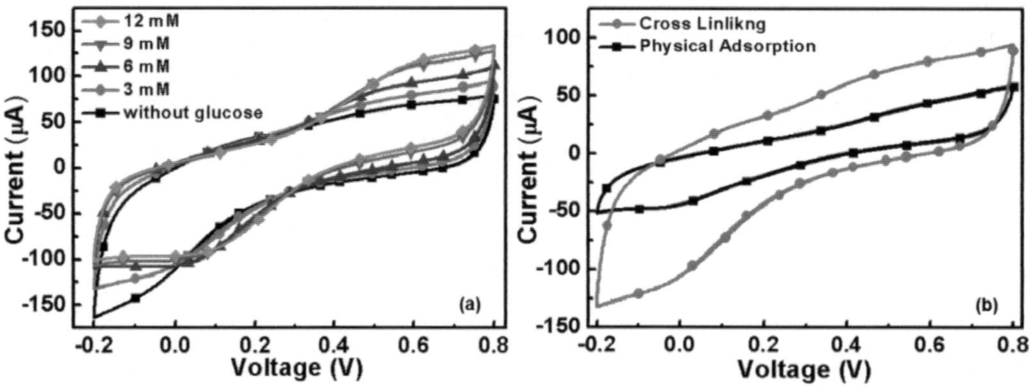

Figure 4. (a) CV obtained by successive injection of glucose for cross linked GO_x, (b) Comparison of CV for biosensors fabricated via physical adsorption of GO_x and cross linking of GO_x.

3.4.2 Amperometric response

Amperometric detection of glucose was used because of its simplicity and accuracy. Current response measurement was performed for the two types of electrodes which were prepared by employing different immobilization techniques. In the catalysis mediated by GO_x glucose gets oxidized into gluconic acid and the enzyme itself gets reduced. Each GO_x molecule has an active site i.e. Flavine adenine dinucleotide (FAD) buried into its protein shell which gets reduced into $FADH_2$ during oxidation of glucose. The enzyme is re-oxidized by the molecular oxygen resulting in evolution of H_2O_2. The released H_2O_2 is quantitatively detected by application of potential to the working electrode as shown in Eqs. (1) - (3). Thus, oxidation of H_2O_2 by applied potential leads to flow of current in the electrochemical cell proportional to the glucose concentration [31].

$$C_6H_{12}O_6 + GO_X(oxd) + H_2O \rightarrow C_6H_{12}O_7 + GO_x(red) \quad (1)$$

$$GO_x(red) + O_2 \rightarrow GO_x(oxd) + H_2O_2 \quad (2)$$

$$H_2O_2 \xrightarrow{0.8V} 2H^+ + 2e^- + O_2 \quad (3)$$

Current density increases on injection of glucose due to generation of electrons. After this for each response calibration curve was obtained which relates the current density to the amount of glucose concentration injected into electrolyte solution. It is hyperbolic in nature which mainly consists of two parts i.e. linear and saturation region. Initially active sites of enzymes were unoccupied and an instantaneous rise in current on addition of glucose results in linear region of calibration curve. While slowly all the active sites were occupied by glucose molecules eventually heading towards saturation with the increase in glucose concentration in the solution. No further increment in the current was observed upon injection of glucose into the electrolyte solution, once saturation was reached.

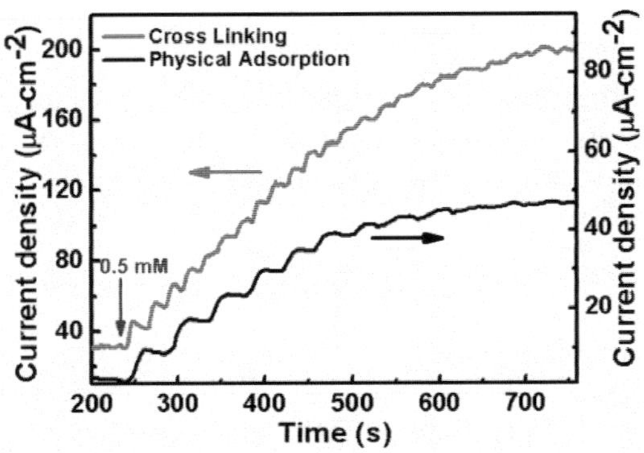

Figure 5. Amperometric response showing current density variation *viz* time for biosensors fabricated via physical adsorption of GO_x and cross linking of GO_x.

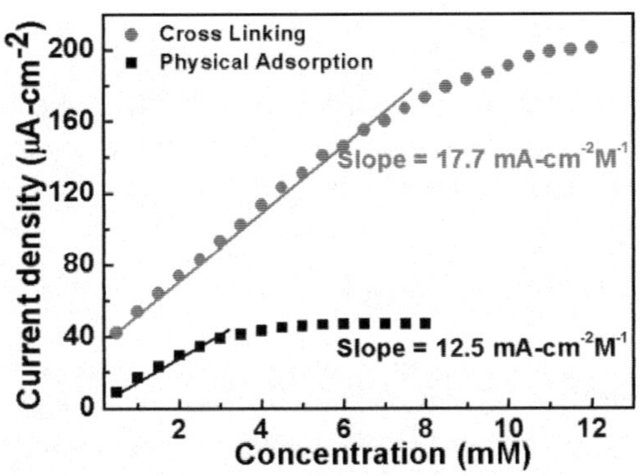

Figure 6. Calibration curves corresponding to as prepared biosensors.

The typical amperometric responses were obtained at an applied potential of 0.8 V for the two types of electrodes prepared by physical adsorption and cross linking of enzymes. The electrolyte solution was continuously stirred to provide convective transport during electrochemical measurement. Figure 5 shows the typical amperometric responses obtained for the same. 0.5 mM glucose has been introduced into the cell at every 25 s for both the electrodes. It has been clearly observed that upon successive injection of glucose current density increases very quickly and saturates resulting in a clear step like response which corresponds to the typical enzyme kinetics behavior as evident from Fig. 5. This indicates that enzymes have been immobilized properly with retained activity over ZNR support matrix. Response time has been defined as the time required by current density to increase by 90 % after addition of glucose [31]. Fast and immediate rise was observed with response time to be around 5-6 s for the two. Moreover calibration curves for two electrodes have been compared and shown in Fig. 6. Sensitivity and linear range was computed from the slope of calibration curves obtained from amperometric responses. Linear range of operation was calculated by linear regression analysis of calibration curves. Sensitivity was calculated to be 12.5 $mA\text{-}cm^{-2}\text{-}mM^{-1}$ with a poor linear range of operation of 0.5-3 mM for electrodes prepared using physical adsorption mode of immobilization. In contrast, sensitivity was enhanced with a wider linear range of operation for cross linking immobilization method which was computed to be 17.7 $mA\text{-}cm^{-2}\text{-}mM^{-1}$ and 0.5 – 8.5 mM respectively. This improvement in figure of merits of biosensor has been attributed to increased enzyme loading due to cross linking mode of immobilization. Cross linking increases amount of enzyme by covalent bond formation with cross linking agent. Table 1 illustrates performance of the two types of biosensors examined.

Table 1. Comparison of the two types of electrodes

Electrode	Sensitivity ($mA\text{-}cm^{-2}\text{-}M^{-1}$)	Response Time (s)	Linear Range (mM)
Physically adsorbed GO_x	12.5	5-6	0.5-3
Cross linked GO_x	17.7	5-6	0.5-8.5

3.4.3 Selectivity

Selectivity of a sensor has been defined as the ability of a sensor to detect a specific analyte of interest. High selectivity is necessary for accurate measurement of analyte. Enzymatic reaction has been widely described by lock and key model in literature and thus enzymes are highly specific towards the substrate [6, 7]. Due to this, it has been utilized widely for biosensing applications.

3.4.4 Interference Study

The human serum contains many electroactive species other than glucose such as Uric acid, Ascorbic acid etc. which may interfere with the biosensor performance while detecting glucose. Therefore it is necessary to evaluate anti interference ability of biosensor. Here anti interference capability of as fabricated cross linked GO_x biosensor was examined by initially injecting 5 mM glucose in the electrochemical cell setup followed by addition of 0.4 mM Uric acid and 0.1 mM Ascorbic acid sequentially. The amount of the three species added into the cell was kept same as the physiological concentration of it in the human body. Figure 7 shows the corresponding amperometric current response of the electrode. As visible from Fig. 7 biosensor almost immediately acquired the steady state after glucose injection while it does not respond to uric acid addition and 4.2 % increment with respect to the current density for 5 mM concentration of glucose was observed for ascorbic acid addition. This anti interference capability of biosensor can be attributed to selectivity of enzymes and presence of nafion as a semipermeable membrane.

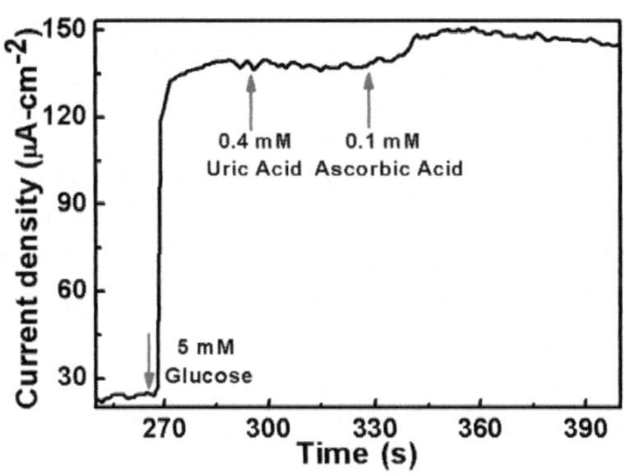

Figure 7. Interference effect study for cross linked GO_x biosensor

4. Conclusions

In summary a simple method involving in situ addition of $KMnO_4$ during hydrothermal growth process was utilized to cease formation of dumbbells in nutrient solution and for growth of ZNR. Highly crystalline ZNR with largely suppressed defects was obtained for studying the biosensor performance. The two types of immobilization techniques *viz* physical adsorption and cross linking have been analyzed and compared. The biosensor figures of merit were found to improve by incorporation of cross linking as immobilization method. Sensitivity of biosensor has increased from 12.5 $mA\text{-}cm^{-2}\text{-}M^{-1}$ for physically adsorbed GO_x to 17.7 $mA\text{-}cm^{-2}\text{-}M^{-1}$ for cross linked GO_x. Apart from this linear range has enhanced significantly from 0.5-3 mM to 0.5-8.5 mM for physically absorbed and cross linked GO_x respectively. Since the support matrix (ZNR) utilized was same for both the electrodes response time, was determined to be equal i.e. 5-6 s. The electrode fabricated employing cross linking mode of immobilization has been found to outperform over physically adsorbed GO_x electrode in terms of sensitivity and linear range along with good anti interference capability. Further improvement in aspect ratio of ZNR can be utilized for achieving even better figure of merits for the as prepared biosensor.

Acknowledgements

One of the authors M. S. is grateful to FESEM, XRD, PL, Fluorescence and Potentiostat/Galvanostat facilities equipped at the Sophisticated Instrument Centre, IIT Indore. M. S. would also like to thank Dr. Mukul Gupta (University Grants Commission Department of Atomic Energy (UGC DAE) Consortium for Scientific Research Indore (M. P.), India) for the usage of the DC magnetron sputtering system. M. S. would further like to thank the Ministry of Human Resource and Development (MHRD), India for providing the Teaching Assistantship (TA). Author, V. S. would like to thank Director of IIT Indore for providing his constant support and encouragement for research.

References

[1] Arya S K, Saha S, Ramirez-Vick J E, Gupta V, Bhansalid S and Singhe S P, Recent advances in ZnO nanostructures and thin films for biosensor applications:Review 2012 Anal. Chim. Acta 737 1-21.
[2] Palod P A and Singh V, Improvement in glucose biosensing response of electrochemically grown polypyrrole nanotubes by incorporating crosslinked glucose oxidase (2015) Mater. Sci. Eng. C 55 420-430.
[3] Fang B, Zhanga C, Wanga G, Wanga M and Jia Y, A glucose oxidase immobilization platform for glucose biosensor using ZnO hollow nanospheres, (2011) Sens. Actuators B 155 304-310.
[4] Singh A K, Dwivedi A D D, Chakrabarti P and Prakash R, Electronic and optical properties of electrochemically polymerized polycarbazole/aluminum Schottky diodes (2009) J. Appl. Phys. 105 1-6.
[5] Solanki P R, Kaushik A, Agrawal V V and Malhotra B D, Nanostructured metal oxide based biosensors, NPG Asia Mater., 3 (2011) 17-24.
[6] Shiryaev M A, Eremin S A and Baranov A N, Biosenors based on zinc oxide 2014 Nanotechnologies in Russia 9 1-17.
[7] Wei A, Pan L and Huang W, Recent progress in the ZnO nanostructure-based sensors 2011 Mater. Sci. Eng. B 176 1409–21.
[8] Bai D, Zhang Z and Yu K, Synthesis, field emission and glucose-sensing characteristics of nanostructural ZnO on free-standing carbon nanotubes films (2010) Appl. Surf. Sci. 256 2643–48.
[9] Janotti A and Van de walle C G, Fundamentals of Zinc Oxide as a semiconductor (2009) Rep. Prog. Phys. 72 1-29.
[10] Rahman M M, Ahammad A J S, Jin J H, Ahn S J and Lee J J, A comprehensive review of glucose biosensors based on nanostructured metal oxides (2010) Sensors 10 4855-86.
[11] Wang Z L, Nanostructures of zinc oxide (2004) Materials today 7 26-33.
[12] Tian Z R, Voigt J A, Liu J, Mckenzie B, Mcdermott M J, Rodriguez M A, Konishi H and Xu H, Complex and oriented ZnO nanostructures (2003) Nat. Mater. 2 821-826.
[13] Kong T, Chan Y, Ye Y, Zhang K, Wang Z and Wang X, An amperometric glucose biosensor based on the immobilization of glucose oxidase on the ZnO nanotubes (2009) Sens. Actuators B, 138 344-350.
[14] Singh S P, Arya S K, Pandey P, Malhotra B D, Saha S, Sreenivas K and Gupta V, Cholesterol biosensor based on rf sputtered zinc oxide nanoporous thin film 2007 Appl. Phys. Lett. 91 1-3.
[15] Hong H, Shi J, Yang Y, Zhang Y, Engle J W, Nickles R J, Wang X and Cai W, Cancer Targeted Optical Imaging with Fluorescent Zinc Oxide Nanowires 2011 Nano Lett. 11 3744-50.
[16] Ahmad M, Pan C, Luo Z and Zhu J, A Single ZnO Nanofiber-Based Highly Sensitive Amperometric Glucose Biosensor (2010) J. Phys. Chem. C 114 9308–13.
[17] Liu J, Guo C, Li C M, Li Y, Chi Q, Huang X, Liao L and Yu T, Carbon-decorated ZnO nanowire array: A novel platform for direct electrochemistry of enzymes and biosensing applications, Electrochem. Commun. 11 (2009) 202–205.

IOP Conf. Series: Journal of Physics: Conf. Series **924** (2017) 012013 doi:10.1088/1742-6596/924/1/012013

[18] Sassolas A, Blum L J and Leca-Bouvier B D, Immobilization strategies to develop enzymatic biosensors 2012 Biotech. Adv. 30 489–511.
[19] Saha S, Tomar M and Gupta V, Influence of stress in ZnO thin films on its biosensing application 2015 Enzyme Microb. Technol. 79 63-69
[20] Sarkar N K and Bhattacharyya S K, High electro-catalytic activities of glucose oxidase embedded one-dimensional ZnO nanostructures 2013 Nanotechnol. 24 1-8
[21] Wei Y, Li Y, Liu X, Xian Y, Shi G and Jin L, ZnO nanorods/Au hybrid nanocomposites for glucose biosensor 2010 Biosens Bioelectron. 26 275–278.
[22] Palod P A and Singh V, Facile synthesis of high density polypyrrole nanofiber network with controllable diameters by one step template free electropolymerization for biosensing applications 2015 Sens. Actuators B 209 85–93.
[23] Dixit T, Bilgaiyan A, Palani I A, Nakamura D, Okada T and Singh V, Influence of potassium permanganate on the anisotropic growth and enhanced UV emission of ZnO nanostructures using hydrothermal process for optoelectronic applications, 2015 J. Sol-Gel Sci. Technol. 75 693–702.
[24] Zhai Y, Zhai S, Chen G, Zhang K, Yue Q, Wanga L, Liu J and Jia J, Effects of morphology of nanostructured ZnO on direct electrochemistry and biosensing properties of glucose oxidase 2011 J. Electroanal. Chem. 656 198–205.
[25] Yu Q, Yu C, Yang H, Fu W, Chang L, Xu J, Wei R, Li H, Zhu H, Li M and Zou G, Growth of Dumbbell-like ZnO Microcrystals under Mild Conditions and their Photoluminescence Properties 2007 Inorg. Chem. 46 6204-10.
[26] Djurisic A B, Leung Y H, Tam K H, Su Y F H, Ding L, Ge W K, Zhong Y C, Wong K S, Chan W K, Tam H L, Cheah K W, Wok W M K and Phillips D L, Defect emissions in ZnO nanostructures 2007 Nanotechnology 18 1-7.
[27] Dixit T, Shukla M, Palani I A and Singh V, Insight of dipole surface plasmon mediated optoelectronic property tuning of ZnO thin films using Au 2016 Opt. Mater. 62 673-679.
[28] Bilgaiyan A, Dixit T, Kapil G, Pandey S S, Hayase S, Palani I A and Singh V, Effect of Addition of KI on the Hydrothermal Growth of ZnO Nanostructures Towards Hybrid Optoelectronic Device Applications 2016 J. Nanosci. Nanotechnol. 16 3301-04.
[29] Yang Y H, Chen X Y, Feng Y and Yang G W, Physical Mechanism of Blue-Shift of UV Luminescence of a Single Pencil-Like ZnO Nanowire 2007 Nano Lett. 7 3879-83.
[30] Dixit T, Palani I A and Singh V, Selective tuning of enhancement in near band edge emission in hydrothermally grown ZnO nanorods coated with gold 2016 J. Lumin. 170 180–186.
[31] Palod P A, Pandey S S, Hayase S and Singh V, Template-Assisted Electrochemical Growth of Polypyrrole Nanotubes for Development of High Sensitivity Glucose Biosensor, 2014 Appl. Biochem. Biotechnol. 174 1059–72.

IOP Conf. Series: Journal of Physics: Conf. Series **924** (2017) 012014 doi:10.1088/1742-6596/924/1/012014

Casting Control of Floating-films into Ribbon-shape Structure by modified Dynamic FTM

A Tripathi[1*], M Pandey[1], S Nagamatsu[2, 3], S S Pandey[1*], S Hayase[1,2] and W Takashima[1]

[1]Graduate School of Life Science and Systems Engineering, Kyushu Institute of Technology, Wakamatsu, Kitakyushu 808-0196, Japan
[2]Reseach Center for Advanced Eco-fitting Technology, Kyushu Institute of Technology, Wakamatsu, Kitakyushu 808-0196, Japan
[3]Department of Computer Science and Electronics, Kyushu Institute of Technology, 680-4 Kawazu, Iizuka, 820-8502, Japan

E-mail: tripathi-atul@edu.life.kyutech.ac.jp / shyam@life.kyutech.ac.jp

Abstract. We have developed a new method to obtain Ribbon-shaped floating films via dynamic casting of floating-film and transfer method (dynamic-FTM). Dynamic-FTM is a unique method to prepare oriented thin-film of conjugated polymers (CPs) which is quick and easy. This method has several advantages as compared to the other conventional casting procedure to prepare oriented CP films. In the conventional dynamic FTM appearance of large scale circular orientation poses difficulty not only for practical applications but also hinders the detailed analysis of the orientation mechanism. In this present work, pros and cons of this newly proposed ribbon-shaped floating-film have been discussed in detail from those of the conventional floating-film prepared by dynamic-FTM.

1. Introduction

Conjugated polymers (CPs) have captivated a lot of attentions in the recent past for their potential applications in organic field effect transistors (OFETs), light emitting diodes (LEDs) and photovoltaics. The solution processing capability of CPs is one of the important features rendering their suitability towards potential application in the area of plastic electronics owing to the cost effective process for fabrication of thin films. Thin film morphology of the CPs plays a dominant role in deciding the charge transport properties, which arises from the self-assembly promoted aggregation of macromolecules while solid phase condensation. Casting procedures to prepare semiconducting layer with CPs is, therefore, an important and decisive step to characterize the transport performance. Although spin-coating is one of the most widely used techniques for fabricating thin films, the random spatial arrangement of main-polymeric chains limits its usage for fabricating high performance organic electronic devices [1]. One of the amicable solutions for this problem is the introduction of macromolecular alignment of CPs. Various methods to prepare oriented thin-films such as friction transfer method [2], mechanical rubbing [3], solution flow [4], capillary action [5], solution shearing [6], slide coating [7] and strain alignment [8] etc. have been developed and investigated in-detail towards the analysis of their orientation characteristics. Although these techniques have demonstrated a potential importance for fabricating organic electronic devices by using oriented thin films of CPs in the last one decade, important challenges like simplicity of fabrication process and preparation of multilayer oriented films are yet to be taken into consideration.

We have developed a dynamic casting of thin floating-film of CPs on liquid-substrate followed by its transfer on a desired substrate which has been named as dynamic-FTM. This is a simple, cheap and quick casting method to provide oriented films of CPs. The oriented area reaches up to centimeter-scale which enables us to construct multilayered thin-films with the preserved oriented morphology [9-11]. However, remaining obstacles like non-uniformity in thickness and control of orientation direction are still remaining challenges which have to be solved. In order to resolve these problems, we have developed a new casting method by modifying the casting of floating films into the ribbon-shape by introducing a custom-made slider to control the spreading of the oriented films. In this study, we have compared the film characteristics prepared by this ribbon-shaped dynamic-FTM and conventional dynamic-FTM. Role of individual casting parameters such as solution concentration, viscosity of liquid-substrate and casting temperature have been investigated by polarized electronic absorption spectroscopy.

2. Experimental

Non-regiocontrolled poly (3-hexyltiophene) (NR-P3HT) was selected as the representative polymeric material in this study. It has been synthesized by chemical oxidative polymerization using $FeCl_3$ catalyst and purified as per our earlier publication [15]. The regioregularity (head-to-tail coupling content) was confirmed to be about 80 % by proton nuclear magnetic resonance investigations. Dehydrated chloroform, purchased from Wako Chemicals, was used for the thin film casting. Synthesized NR-P3HT was highly soluble in the dehydrated chloroform and used to prepare two types of solutions, 1% and 2% (wt/wt). Procedure to cast the ribbon shaped floating-film has been schematically shown in the Fig. 1. Hydrophilic liquid mixture was first made and poured into a rectangular tray which serves the purpose of liquid-substrate. A hand-made slider made of PTFE was put near the one of edges in the tray. About 15 μl of NR-P3HT solution in chloroform was dropped to the slider. When the droplet reached to the slider/liquid-substrate boundary, there was a quick and unidirectional spreading of the solution followed by continuous solidification leading to the fabrication of ribbon-shaped floating-film. After this casting, the obtained floating-film was left for 10 minutes to ensure complete evaporation of the solvent. Since the orientation of CP in the floating-film was of macroscopic in order, it was easily confirmed through a linearly polarized film with the naked eyes.

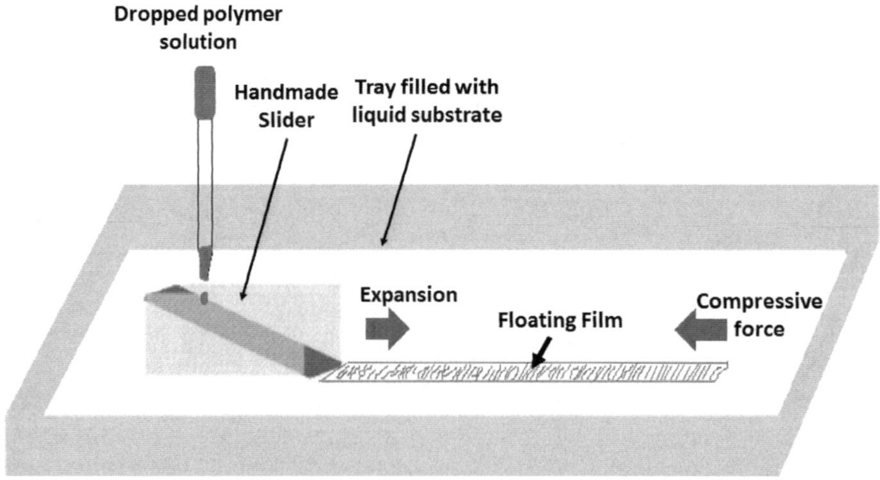

Figure 1. Schematic representation of the set-up for casting ribbon-shaped floating-film

The resulted floating film was stamped on transparent solid-substrates for the orientation analysis. The surface of the solid substrates were treated with hexamethyldisilazane (HMDS) to increase the hydrophobicity in order to ensure easy and better adhesion of the thin floating-films during stamping. The transferred film surface was washed with methanol in order to remove any remaining liquid substrate on the stamped surface followed by drying. In order to investigate the effect of liquid-substrate viscosity on the molecular orientation, a variety of binary solvent mixtures like water and ethylene glycol (Wt/Eg) or ethylene glycol and glycerol (Eg/Gl) were used to prepare the liquid-substrates as per our earlier publications 14,15]. Optical anisotropy in the oriented thin films was investigated by the polarized electronic absorption spectra obtained through a Glan-Thompson prism with JASCO V-570 spectrophotometer. The orientation intensity of the NR-P3HT in these films was calculated from polarized absorption spectra as the optical dichroic ratio (DR).

3. Results and Discussion

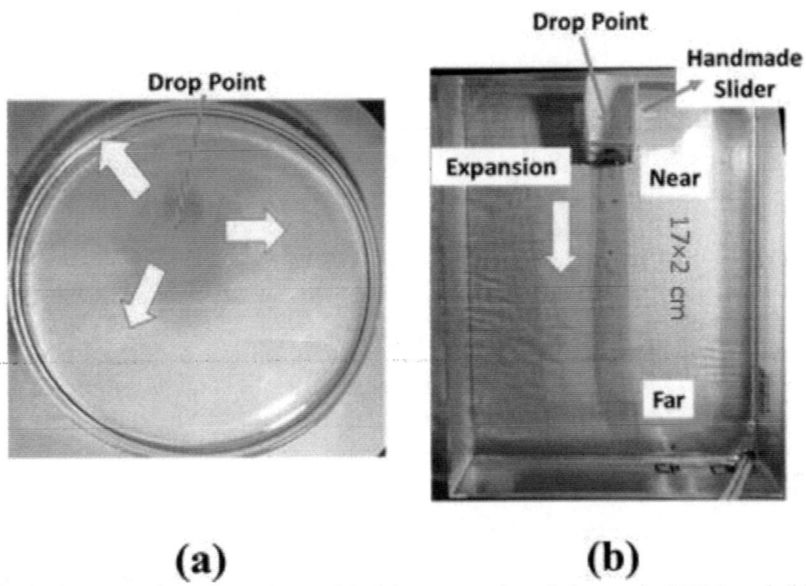

Figure 2. Optical photographs of casting with (a) conventional dynamic FTM and (b) Ribbon-shape FTM procedure, respectively.

Figure 2 shows the photographic images of the obtained floating-films prepared by the conventional and Ribbon-shaped FTM methods, respectively. It is clearly observed that the obtained shape of the floating-film drastically changes depending on the employed method of the film casting under dynamic FTM. A circular floating-film with some dispersive parts formed in the floating film can be clearly seen for the thin film fabrication using conventional dynamic-FTM. On the other hand, this newly introduced slider changed the shape of the floating-film into the Ribbon-shaped one. It is obvious that the later method has advantages in terms of detailed analysis as well as for the easy handling the transfer process due to its large scale rectangular shape. Polarized electronic absorption spectra of Ribbon-shaped floating-film casted on glass substrate has been shown in the Fig.3. A non-polarized electronic absorption spectra of a spin-coated NR-P3HT film has been also incorporated in order to compare the orientation behavior. DR of the Ribbon-shaped floating-films was found to be 2.4. It can be easily seen that absorption maximum (λ_{max}) in the electronic absorption spectrum of parallel oriented NR-P3HT film was located at 505 nm along with the presence of vibronic shoulders. Interestingly, λ_{max} in the electronic absorption spectrum of spin-coated film was found to be at 498 nm with a clear red shift of 7 nm in the Ribbon-shaped FTM film. Vibronic shoulders around 540 nm and 602 were appeared in the similar manner to that observed for the conventional dynamic-FTM.

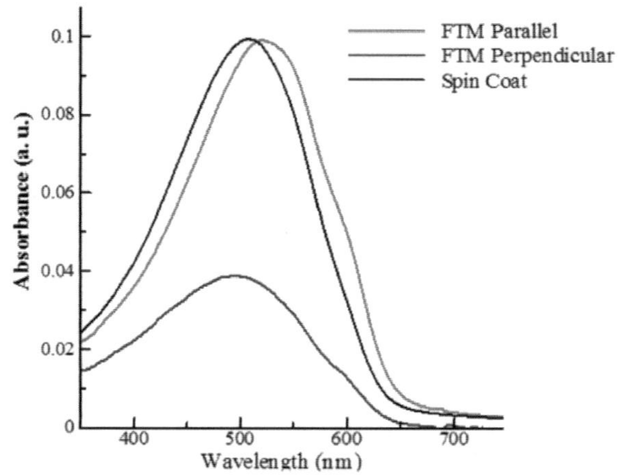

Figure 3. Polarized electronic absorption spectra of NR-P3HT thin films prepared by spin coating and Ribbon-shape FTM

These findings represent the presence of the increased π-orbital delocalization on main-chain [16]. The results support that the films fabricated by this newly developed method also have enhanced π-stacking and such films are suitable for the fabrication of optoelectronic devices. A relatively small DR of 2.4 observed in this ribbon-shaped oriented film indicates that developed slider method provides a weak orientation as compared to that of the conventional dynamic-FTM. A clear reason for this cannot be assigned at the present stage but further development and optimization of the slider shape is required for achieving the high orientation. Although the orientation intensity is small in the ribbon-shaped films, it is worthy to note that λ_{max} is clearly red-shifted even in NR-P3HT as discussed above. It has been already demonstrated, that thin film casted via dynamic-FTM is found to show high transport performance as compared to that of the spin-coat film [12]. In addition, the orientation size can be reached up to several centimeter-scales and better than the conventional dynamic-FTM.

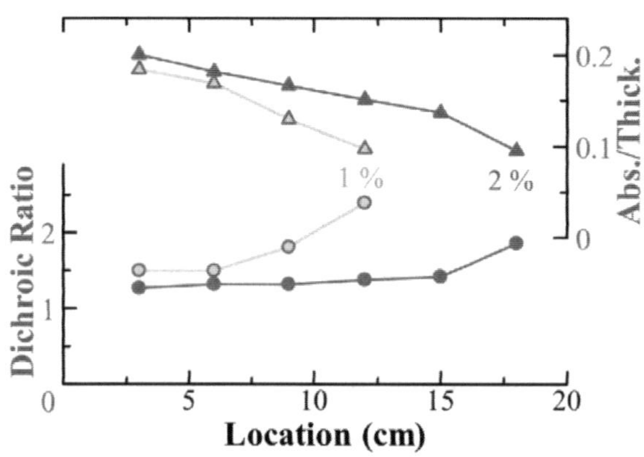

Figure 4. Variation of DR and peak absorbance with respect to the casting point (near). Variation in peak absorbance represents that the thickness of the film is less in the far region and higher towards the beginning (slider).

Figure 4 shows the location dependence of DR as well as the absorbance at the λ_{max}. Absorbance at λ_{max} at different locations of the ribbon-shaped FTM was also measured in order to investigate variation in thickness of the films. It is important to note that thin films casted towards the slider end in the Ribbon-shaped FTM exhibit nearly no change in the measured DR values up to >10 cm. This indicates that the

Ribbon-shaped casting method provides relatively uniform orientation especially towards the slider side. The film thickness tends to be thinner as a function of increasing distance of the casted film from slider side to opposite end. These location dependences were also observed even after changing the concentration of the polymer solution. Such information about the distribution of large-scale orientation is highly desired for the in-depth discussion about the orientation mechanism. In this context, Ribbon-shaped thin film casting procedure proposed here provides the considerable knowledge of orientation characteristics by dynamic-FTM. It is interesting to note that there is appreciable enhancement in the observed DR at the far area in the Ribbon-shaped floating-film. This could be explained considering the fact that at the initial stage of solidification of floating film, extent of molecular orientation is less due to the fast solvent evaporation and dragging viscous force from the liquid substrate. There is induction of molecular orientation is the result of competition between the film spreading and opposing viscous force posed by the liquid substrate which continues further with the advancing floating film.

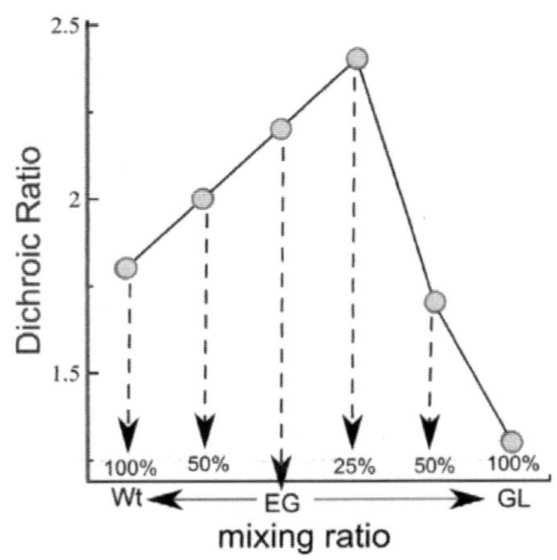

Figure 5. DR dependence on the binary liquid mixture of water, EG or GL as casting liquid-substrate

Figure 5 depicts the liquid substrate viscosity dependence of observed DR in the Ribbon-shaped floating-film where the viscosity was controlled by mixing the two miscible hydrophilic solvents of varying viscosities. All of the DR plotted here are taken at the farthest position giving the best orientation. Although DR is small as mentioned above, the orientation characteristics are quite the same of the floating-films obtained by the conventional dynamic-FTM as reported previously [14]. This fact also indicates that even by changing floating film fabrication method as mentioned above with Fig.2, the essential orientation characteristics remain well conserved. This is also very important to note that the Ribbon-shaped dynamic-FTM is useful for investigating the mechanism with various casting factors aiming towards the attainment of high molecular orientation. The proposed Ribbon-shaped dynamic-FTM is utilized for researching the orientation mechanism in dynamic-FTM and for optimizing the orientation condition towards the practical use.

4. Conclusions

We have developed a modified dynamic-FTM to obtain Ribbon-shaped floating-films. The resulting floating-films were confirmed to be oriented to perpendicular direction with respect to the propagation direction of the floating film. The orientation intensity given by the dichroic ratio (DR) was 2.4, which was relatively small as compared to the conventional dynamic-FTM. DR dependences on the distance from the dropping point, solution concentration and liquid-substrate were investigated as a preliminary investigation. These results exhibit almost the same characteristics as compared that observed in conventional dynamic FTM. The developed casting way can be utilized for investigating the orientation mechanism as well as for the optimizing the casting condition in dynamic-FTM.

IOP Conf. Series: Journal of Physics: Conf. Series **924** (2017) 012014 doi:10.1088/1742-6596/924/1/012014

Acknowledgments
Authors sincerely express their thanks to Japan Society for the Promotion of Science (JSPS) for the financial support by Grant-in-Aid for Scientific Research (C) (Grant Number 15K05989).

References
[1] Strawhecker K E, Kumar S K, Douglas J F and Karim A 2001 *Macromolecules* **34** 4669–72
[2] Nagamatsu S, Takashima W, Kaneto K, Yoshida Y, Tanigaki N, Yase K and Omote K 2003 *Macromolecules* **36** 5252–7
[3] Hartmann L, Tremel K, Uttiya S, Crossland E, Ludwigs S, Kayunkid N, Vergnat C and Brinkmann M 2011 *Adv. Funct. Mater.* **21** 4047–57
[4] Yamasaki N, Miyake Y, Yoshida H, Fujii A and Ozaki M 2011 *Jpn. J. Appl. Phys.* **50**
[5] Higashi T, Yamasaki N, Utsumi H, Yoshida H, Fujii A and Ozaki M 2011 *Appl. Phys. Express* **4** 91602
[6] Giri G, Delongchamp D M, Reinspach J, Fischer D A, Richter L J, Xu J, Benight S, Ayzner A, He M, Fang L, Xue G, Toney M F and Bao Z 2015 *Chem. Mater.* **27** 2350–9
[7] Karakawa M, Chikamatsu M, Yoshida Y, Oishi M, Azumi R and Yase K 2008 *Appl. Phys. Express* **1** 0618021–3
[8] O'Connor B, Kline R J, Conrad B R, Richter L J, Gundlach D, Toney M F and DeLongchamp D M 2011 *Adv. Funct. Mater.* **21** 3697–705
[9] Morita T, Singh V, Nagamatsu S, Oku S, Takashima W and Kaneto K 2009 *Appl. Phys. Express* **2** 1–4
[10] Morita T, Singh V, Oku S, Nagamatsu S, Takashima W, Hayase S and Kaneto K *2010 Jpn. J. Appl. Phys.* **49** 0416011–5
[11] Noh J, Jeong S and Lee J-Y 2016. *Nat. Commun.* **7** 12374
[12] Pandey M, Nagamatsu S, Pandey S S, Hayase S and Takashima W 2016 *Org. Electron.* **38** 115–20
[13] Dauendorffer A, Miyajima S, Nagamatsu S, Takashima W, Hayase S and Kaneto K 2012 *Appl. Phys. Express* **5**
[14] Pandey M, Pandey S S, Nagamatsu S, Hayase S and Takashima W 2017 *J. Nanosci. Nanotechnol.* **17** 1915–22
[15] Pandey M, Nagamatsu S, Pandey S S, Hayase S and Takashima W 2016 *J. Phys. Conf. Ser.* **704** 12005
[16] Pandey M, Pandey S S, Nagamatsu S, Hayase S and Takashima W 2016 *Thin Solid Films* 619 125–30

IOP Conf. Series: Journal of Physics: Conf. Series **924** (2017) 012015 doi:10.1088/1742-6596/924/1/012015

Oriented antibody immobilization on self-assembled monolayers applied as impedance biosensors

Kaiki Tsugimura[1], Hitoshi Ohnuki[1]*, Haiyun Wu[2], Hideaki Endo[2], Daiju Tsuya[3] and Mitsuru Izumi[1]

[1] Tokyo University of Marine Science and Technology, 2-1-6 Etchujima, Koto-ku, Tokyo 135-8533, Japan

[2] Tokyo University of Marine Science and Technology, 4-5-7 Konan, Minato-ku, Tokyo 108-8477, Japan

[3] National Institute for Materials Science, 1-2-1 Sengen, Tsukuba, Ibaraki 305-0047, Japan

*Correspondence: ohnuki@kaiyodai.ac.jp

Abstract. Oriented immobilization of antibodies on a sensor chip is crucial for enhancing both the sensitivity and antigen-binding capacity of immunosensors. Here, we report a comparative study of the effect of oriented and random antibody immobilization on the binding efficiency by electrochemical impedance spectroscopy (EIS). Oriented immobilization of anti-myoglobin immunoglobulin G (anti-Myo IgG) was achieved by bonding to an Fc receptor of protein G (PrG) on a self-assembled monolayer (SAM), which results in the myoglobin (Myo) binding sites being exposed outside the sensing surface. Random immobilization of anti-Myo IgG was achieved by direct covalent attachment to the SAM surface. Both immobilizations were applied to interdigitated electrodes to enhance the electrochemical signal, and the Myo biosensor performance was then evaluated by a series of EIS measurements. We found that (i) the rate of the normalized charge transfer resistance for the oriented sample was 3 times higher than that for the random sample and (ii) the detection limit was 0.001 ng/mL, which is the lowest recorded detection limit among Myo immunosensors based on EIS. These findings indicate that oriented antibody immobilization is crucial for preparing highly sensitive EIS-based biosensors.

1. Introduction

Immobilization of antibodies not only produces functionalized sensing surfaces but it also determines the sensitivity of immunosensors [1]. Covalent immobilization is the preferred method for antibody immobilization. Owing to their abundance, the amino/carboxylic groups on the antibodies are mostly used for this process. However, antibodies have these groups almost uniformly over their surface, which leads to randomly oriented immobilization with respect to the sensing surface. Such randomness may hinder antibody–antigen binding, resulting in poor sensor performance [2]. This problem can be avoided by using protein A (PrA) or protein G (PrG), which are found on the surface of *Staphylococcus aureus* or *Streptococcus*, for oriented immobilization of the antibody [3]. These proteins specifically bind to the non-antigenic Fc region of the antibody, and the antigen binding sites of the antibody are then exposed outside the sensing surface, leading to better antigen binding capacity [4,5] and decreasing non-specific binding [6] (see figure 1). Compared with PrA, PrG possesses a wide range of binding abilities to immunoglobulin G (IgG) subclasses, which comprise 70–80% of total serum antibodies and have high affinity constants [7]. These features make PrG very attractive as an oriented IgG binding support for immunosensor applications.

Electrochemical impedance spectroscopy (EIS) is a very powerful tool to analyze the interfacial properties related to bio-recognition events occurring at the sensing surface [8,9]. EIS directly analyzes the resistance and capacitance of the surface. Both parameters are sensitive to the occurrence of molecular adsorption, such as Myo and anti-Myo IgG binding, without any need for labelling or other pretreatment processes. In most cases, the resistance change is used as the analysis parameter because of its high sensitivity with ease of detection through a Nyquist plot ($-Z_{im}$ vs. $Z_{re,}$ where Z_{im} and Z_{re} are the imaginary and real parts of the complex impedance, respectively). In this study, we used the faradic EIS technique with a $[Fe(CN)_6]^{3-/4-}$ redox probe to detect the resistance, because this technique has received much attention for application to immunodetection [10], aptamer-based protein biosensing [11], and genosensing [12]. To enhance the EIS signal, we used a patterned substrate of interdigitated electrodes (IDEs) composed of a series of parallel microbands electrodes with alternating microbands connected together [13]. Because of the short distance between the anode and cathode electrodes, the speed of the oxidation and reduction cycle of the redox species is greatly enhanced, which results in highly efficient data collection [14].

In this work, we developed a label-free Myo biosensor based on the affinity binding between Myo and anti-Myo IgG. The sensing target Myo is the earliest marker molecule of acute myocardial infarction (AMI). The level of Myo significantly increases as early as 1 h after the onset of AMI and peaks within 3–6 h [15]. These characteristics facilitate early confirmation of AMI and help in clinical decision making. The oriented anti-Myo IgG surface was prepared by attaching anti-Myo IgG to a PrG layer covalently immobilized on a self-assembled monolayer (SAM). Random anti-Myo IgG was obtained by direct covalent attachment to the SAM surface. The effect of the IgG orientation and the Myo sensing performance were evaluated by EIS through a comparative study of oriented and random IgG immobilization surfaces.

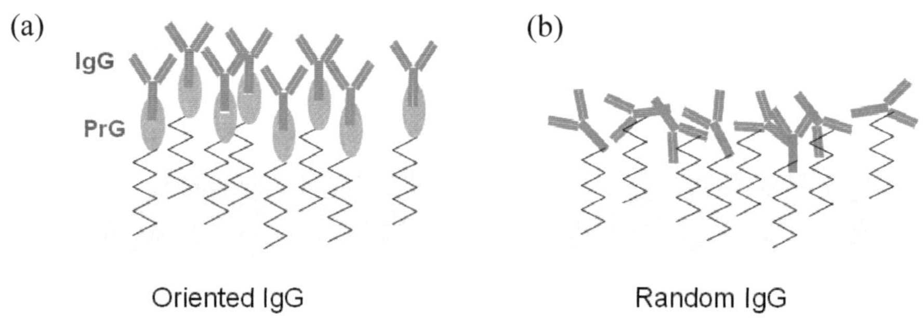

Figure 1. Schematic diagrams of IgG immobilization on the electrode surface: (a) oriented IgG immobilization via PrG and (b) random immobilization with covalent attachment to the SAM surface.

2. Materials and methods

2.1. Materials

11-Mercaptoundecanoic acid (MUA), 6-mercaptohexanol (C6OH), bovine serum albumin (BSA), phosphate-buffered saline (PBS, pH 7.2), recombinant PrG, myoglobin from horse skeletal muscle (Myo), 1-ethyl-3-(3-dimethylaminopropyl)carbodiimide hydrochloride (EDC), and *N*-hydroxysuccinimide (NHS) were purchased from Sigma-Aldrich Japan (Tokyo, Japan). Recombinant PrG was used because it shows better sensing performance than native PrG [16]. Anti-Myo IgG was purchased from Bethyl (Montgomery, TX, USA). The IDEs were fabricated on a quartz glass plate by photolithography techniques. Ti/Au (40/200 nm) layers were deposited on the glass plate. IDEs with

52 fingers with length 8.0 mm and width 50 μm with a 50 μm gap between adjacent fingers were patterned by the lift-off process.

2.2. Preparation of the mixed SAM

The IDE substrate was pretreated in piranha solution (3:1 mixture of H_2SO_4 and H_2O_2) for 10 min and rinsed with deionized water. The substrate was then immersed in a binary mixture of MUA and C6OH ethanolic solution with a concentration of 1.0 mM + 9.0 mM (10% MUA) (figure 2). The immersion period was 24 h and the samples were thoroughly rinsed with ethanol and deionized water, and then dried under N_2 flow. It should be noted that in a previous our study, the use of mixed SAMs with MUA and C_6OH was shown to improve the EIS-based biosensing performance [17].

2.3. Oriented immobilization of anti-Myo IgG

To immobilize the PrG layer on the mixed SAM through covalent bonding, the carboxyl group of MUA was activated in an aqueous solution containing 200 mM EDC and 50 mM NHS for 15 min. After washing with deionized water, the sample was exposed to 0.1 mg/mL PrG solution in PBS for 20 min. Following this procedure, PrG molecules were immobilized on the IDE surface by covalent bonds. Subsequently, unbound PrG was removed by washing with PBS. Finally, the IDE was immersed in PBS containing 1.5% BSA for 1 h to block the non-specific sites and then washed twice with PBS and deionized water. Freshly prepared 0.1 μg/mL anti-Myo IgG solution in PBS was applied to the PrG-modified IDE surface for 2 h. The electrodes were then thoroughly washed with PBS and deionized water to remove any unbound IgG fragments and dried under a N_2 stream.

2.4. Random immobilization of anti-Myo IgG

Random immobilization of IgG was performed by direct covalent attachment to the mixed SAM surface. Similar to the procedure for oriented immobilization (section 2.3), the carboxyl group on the mixed SAM was activated in a mixed solution of 200 mM EDC and 50 mM NHS for 2 h. After washing with deionized water, the sample was immersed in 0.1 μg/mL anti-Myo IgG solution for 2 h and washed with deionized water. Finally, the residual NHS esters were blocked with 1.5% BSA solution in PBS for 1 h. The electrode was washed with PBS and deionized water, and dried under a N_2 stream.

2.5. Measurements and apparatus

To characterize the Myo sensor performance, the prepared samples were immersed in PBS containing different concentrations of Myo at 25 °C for 30 min and carefully washed with PBS. EIS measurements were performed in PBS containing a mixture of 5 mM $K_4[Fe(CN)_6]$ and 5 mM $K_3[Fe(CN)_6]$. The EIS spectra were recorded in the frequency range of 0.1 to 1×10^5 Hz with a 10 mV amplitude using an AutoLab PGSTAT128N potentiostat (Metrohm, Netherlands).

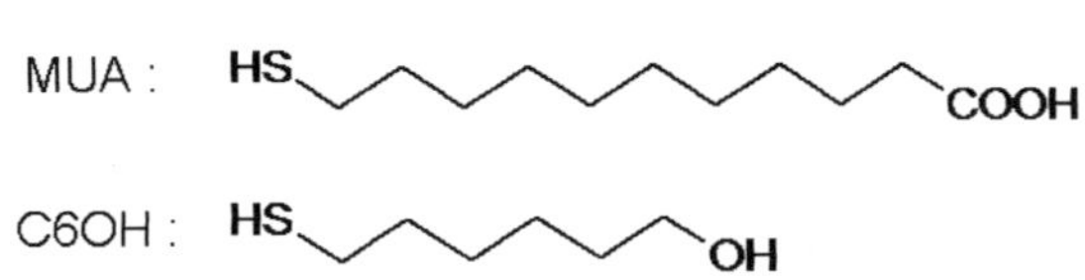

Figure 2. Structural formulas of MUA and C6OH that form the mixed SAM.

3. Results and discussion

3.1. Results of EIS measurements

To characterize the sensor performance, successive cycles of immersion in a fixed Myo concentration and corresponding EIS measurements were performed for each Myo concentration from the lowest to the highest Myo concentration. Figure 3 shows Nyquist plots of the EIS spectra for the oriented IgG sample exposed to various Myo concentrations (0.001–1000 ng/mL). Each spectrum has a semicircular profile. As the Myo concentration increases, the diameter of the semicircle gradually increases. In general, the diameter of the semicircle represents the interfacial charge transfer resistance R_{ct}, that is, the resistance corresponding to carrier transfer between the IDE and $[Fe(CN)_3]^{3-/4-}$ in solution. Accordingly, the observed diameter increase is attributed to formation of the antigen–antibody complex between Myo and anti-Myo-IgG, where the adsorbed Myo effectively blocks $[Fe(CN)_3]^{3-/4-}$ resulting in an increase in R_{ct}. An increase in R_{ct} is observed in the concentration range 0.001–10 ng/mL. Above this range, very small increases in the diameter are observed for the spectra. We consider that this behaviour corresponds to saturation of the Myo binding sites.

Figure 4 shows the spectra for the random IgG sample. Similar to the oriented sample, each spectrum has a semicircular shape. The diameter increases with increasing Myo concentration, which indicates an increasing amount of Myo bound to the sensor surface. These observed behaviours are very similar to those observed for the oriented IgG sample. However, some different features are found for the random IgG sample. First, the semicircle diameter of the initial surface (0 ng/mL Myo) is three times larger than that for the oriented IgG sample. Such a large initial value of R_{ct} is considered to originate from a large amount of IgG immobilization leading to a highly resistive surface. Second, the relative R_{ct} change with respect to R_{ct} (0 ng/mL) remains small compared with the oriented IgG sample, which will be discussed in detail in section 3.2.

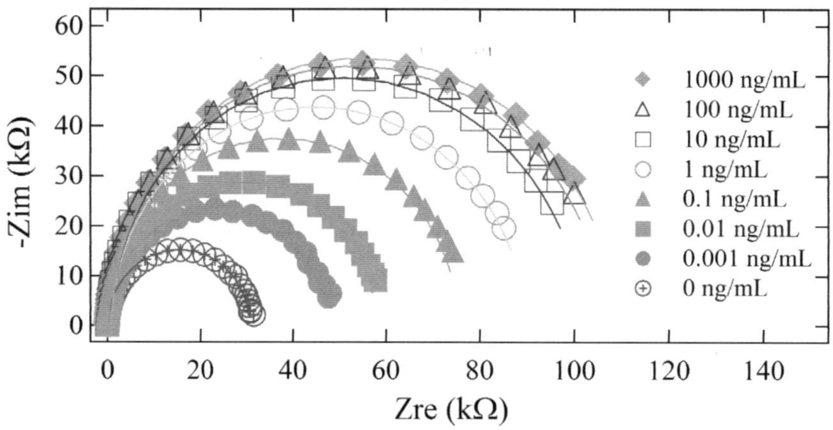

Figure 3. Nyquist plots of the impedance spectra for the oriented anti-Myo-IgG sample immersed in various concentrations of Myo. The solid lines are the fitted curves obtained using the equivalent circuit in figure 5.

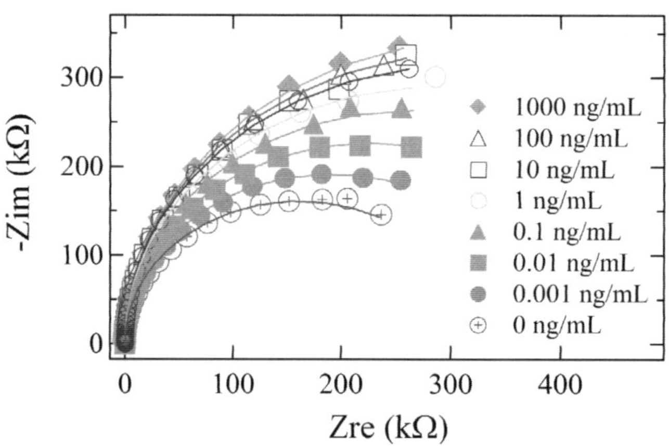

Figure 4. Nyquist plots of the impedance spectra for the random anti-Myo-IgG sample immersed in various concentrations of Myo. The solid lines are the fitted curves obtained using the equivalent circuit in figure 5.

3.2. Equivalent circuit analysis
To appropriately evaluate R_{ct}, the EIS results were simulated using the equivalent circuit shown in figure 5. Here, the solution–electrode interfacial part is expressed as a parallel circuit of R_{ct} and C_{dl} (double-layer capacitance). At low frequencies, where the charge transfer might be faster than the diffusion of the redox substance to the electrode, Warburg impedance Z_W must be introduced in series to R_{ct}. The parameter R_s represents the resistance of the bulk solution. We then performed numerical fittings of all of the spectra, which are shown as solid lines in figures 3 and 4. In figure 6, for both the oriented and random IgG immobilization samples, the calibration curves of calculated R_{ct} versus logarithmic IgG concentration are plotted, where normalization is applied as R_{ct}/R_{ct} (0 ng/mL). The normalized R_{ct} linearly increases with increasing logarithmic IgG concentration for both samples. This linear increase is commonly observed for many affinity biosensors [9] and it is well explained in terms of the Temkin isotherm adsorption model [18]. In figure 6, the important point is that the oriented sample shows a significantly higher rate of increase in R_{ct} compared with the random sample. This behaviour indicates that the oriented sample exhibits better sensor performance than the random sample. For the Myo sensing performance, the oriented sample shows a linear increase of the normalized R_{ct} in the concentration range 0.001–10 ng/mL, and the rate of increase decreases above this range. This shows that the possible Myo detection range for the oriented sample is 0.001–10 ng/mL. Conversely, the random sample shows a linear increase in the range 0.001–1000 ng/mL, although the rate of increase is comparatively small. The observed detection limit of 0.001 ng/mL for both samples is the lowest reported detection limit for an Myo immunosensor based on EIS [19, 20]. We believe that such high sensitivity is the result of signal enhancement by the IDEs.

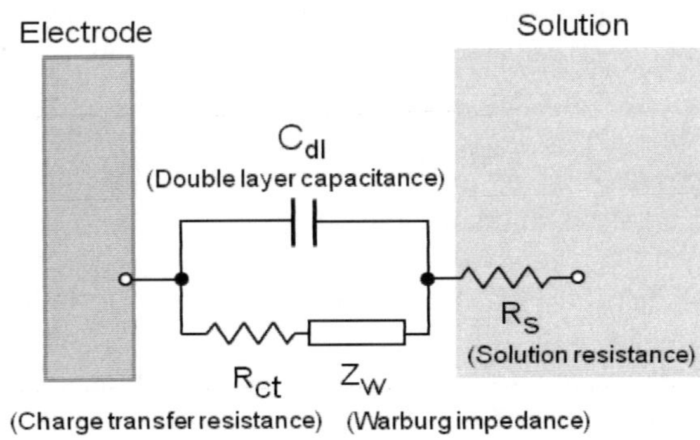

Figure 5. Equivalent circuit model used for the analysis.

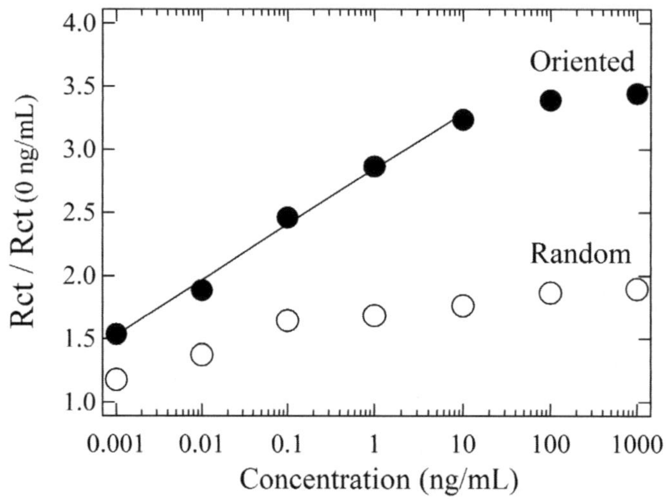

Figure 6. Calibration curves of Myo immunosensors with oriented and random immobilization of anti-Myo IgG on IDEs. The best-fit line for the linear range of the oriented plots is $R_{ct}/R_{ct(0\ ng/mL)} = 0.438 \log C_{Myo} + 2.83$ ($R^2 = 0.997$).

3.3. Effect of oriented IgG immobilization

The high performance of the oriented IgG sample can be explained by the enhanced antigen-binding efficiency. As shown in figure 7, the amount of immobilized IgG on the oriented sample is smaller than that on the random sample. However, for the same sample, a large increase in the rate of normalized R_{ct} with the Myo concentration is observed. This indicates that the oriented anti-Myo IgG sample has higher binding efficiency to Myo than the random sample. This is because most of antigen binding sites are directly exposed to the assay solution, which leads to higher binding efficiency with Myo. In contrast, for the random orientation, although a large amount of IgG is immobilized, there are only a limited number of binding sites exposed to the solution, which results in lower binding efficiency. Similar results of increased binding capacity by oriented antibody immobilization have been reported in previous studies with other experimental techniques [2, 21, 22]. In most of the previous studies, oriented antibody coupling techniques give an antigen-binding capacity that is 2–8 times higher than the efficiencies obtained with the random coupling method. In our study, we found that the oriented anti-Myo IgG sample shows a 3 times higher increase in the rate of normalized R_{ct} than the random sample. Although no previous studies have been reported using the EIS technique for a similar PrG–IgG system, the increase in the rate of normalized R_{ct} in this study is consistent with previous values.

For the Myo sensing performance, we found saturation of anti-Myo IgG binding sites in the high concentration range (>10 ng/mL) for the oriented sample. In contrast, no clear saturation behaviour is observed for the random sample. Even though a slight decrease of the increasing trend occurs at 0.1 ng/mL, the signal continuously increases at higher concentrations. The different saturation behaviours can be explained by different efficiencies of binding. In the oriented IgG configuration, the binding efficiency to the antigen is high and homogeneous for all of the IgG molecules immobilized on the surface, which leads to saturation in the low concentration range. However, in the random orientation configuration, IgG possesses its own binding efficiency with a large variety of the degree depending on the direction, which leads to saturation in a high concentration environment.

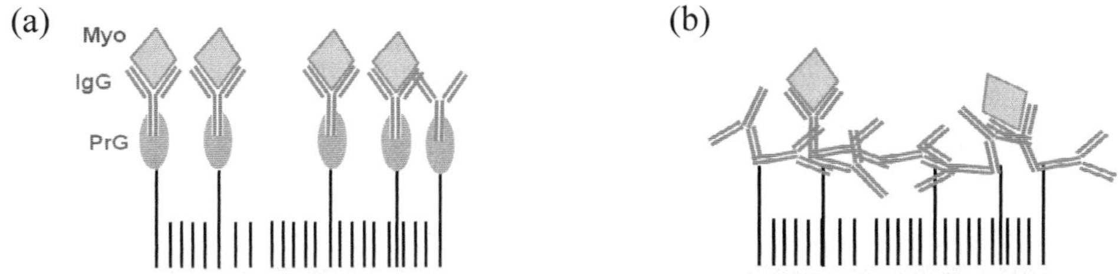

Figure 7. Possible surface models for (a) oriented and (b) random IgG immobilization.

4. Conclusions

We have investigated the effect of the antibody orientation on EIS-based biosensors. Oriented anti-Myo IgG was prepared by immobilizing the F_c receptor PrG on the SAM surface, which resulted in the F_{ab} binding domain pointing away from the surface. Random immobilized anti-Myo IgG was also prepared by direct covalent bonding to the SAM surface as a reference. EIS measurements revealed that the rate of increase of the normalized R_{ct} for the oriented sample is 3 times higher than that for the random sample. In addition, saturation of IgG binding sites is observed at around 10 ng/mL Myo for the oriented sample, but no clear saturation behaviour is observed for the random sample. These findings can be explained by the homogeneous and higher binding efficiency of the oriented IgG surface compared with the random surface. The results suggest that this strategy of orientated IgG immobilization using PrG will be applicable to a variety of EIS-based biosensors.

Acknowledgment

This work was supported by JSPS KAKENHI (Grant Number 25390050). Part of this study was supported by the NIMS Nanofabrication Platform in the Nanotechnology Platform Project sponsored by the Ministry of Education, Culture, Sports, Science and Technology (MEXT), Japan.

References

[1] Willner I and Katz E 2005 *Bioelectronics: From Theory to Applications* (Weinheim: Wiley-VCH) Chap. 6
[2] Neubert H., Jacoby E S, Bansal S S, Iles R K, Cowan D A and Kicman A T 2002 *Anal. Chem.* **74** 3677
[3] Goward C R, Murphy J P, Atkinson T, and Barstow D A 1990 *Biochem. J.* **267** 171
[4] Spitznagel T M and Clark D S 1993 *Nat. Biotechnol.* **11** 825
[5] Tajima N, Takai M and Ishihara K 2001 *Anal. Chem.* **83** 1969
[6] Durovsky T, Tronin A, Dubrovskaya S, Vakula S and Nicolini C 1995 *Sensors and Actuators B* **23** 1
[7] Fahnestock S R 1987 *Trends. Biotechnol.* **5** 79
[8] Daniels J S and Pourmand N 2007 *Electroanalysis* **19** 1239
[9] Katz E and Willner I 2003 *Electroanalysis* **15** 913
[10] Ouerghi O, Senillou A, Jaffrezic-Renault N, Martelet C, Ouada H and Cosnier S 2001 *J. Electroanal. Chem.* **501** 62
[11] Xu D, Xu D, Yu X, Liu Z, He W and Ma Z 2005 *Anal. Chem.* **77** 5107
[12] Liu J, Tian S, Nielsen P and Knoll W 2005 *Chem. Commun.* 2969
[13] Ohno R, Ohnuki H, Wang H, Yokoyama T, Endo H, Tsuya D and Izumi M 2013 *Biosens. Bioelectron.* **40** 422
[14] Niwa O, Morita M and Tabei H 1990 *Anal. Chem.* **62** 447
[15] Sallach S, Nowak R, Hudson M, Tokarski G, Khoury N, Tomlanovich M, Jacobsen G, Lemos J and McCord J 2004 *Am. J. Cardiol.* **94** 864
[16] Matharu Z, Bandodkar A, Gupta V and Malhotra B 2012 *Chem Soc Rev* **41** 1363
[17] Tsugimura K, Ohnuki H, Endo H, Tsuya D and Izumi M 2016 *Jpn. J. Appl. Phys.* **55** 02BE06

[18] Johnson R and Arnold F 1995 *Biochim. Biophys. Acta* **1247** 293
[19] Billah M, Hodges C, Hays H and Millner P 2010 *Bioelectrochemistry* **80** 49
[20] Moreira F, Dutra R., Noronha J and Sales M 2013 *Electrochimica Acta* **107** 481
[21] Makaraviciute A and Ramanaviciene A 2013 *Biosens. Bioelectron.* **50** 460
[22] Lu B, Smyth M and O'Kennedy R 1996 *Analyst* **121** 29R

IOP Conf. Series: Journal of Physics: Conf. Series **924** (2017) 012016 doi:10.1088/1742-6596/924/1/012016

M intermediate accumulation analysis of bacteriorhodopsin reconstituted with three partial peptides

Y Tsujiuchi [1]* H Masumoto [2] and T Goto [3]

[1] Department of Material Science and Engineering, Akita University, 1-1Tegata-Gakuen, Akita, 010-8502 Japan

[2] Frontier Research Institute for Interdisciplinary Sciences, Tohoku University, Sendai 980-8578, Japan

[3] Institute for Materials Research, Tohoku University, 2-1-1 Katahira, Aoba, Sendai, 980-8577 Japan

E-mail: tsuji71@ipc.akita-u.ac.jp

Abstract. Reconstituted bacteriorhodopsin containing helices C-G, A, and B has unique thermal structural change and photoreaction. Time evolution of photo reaction of bacteriorhodopsin reconstituted with three partial peptides in glycerol mixed purple membrane at around 196 K under irradiation by red light was analyzed by a proposed model. The absorption change of 560 nm and 405 nm with irradiation times was analyzed for the purpose of determining reaction rates of photo reaction of reconstituted bacteriorhodopsin and its product M intermediate. In this study it is shown that reaction rates of conversion from bacteriorhodopsin reconstituted with three partial peptides to the M intermediate can be explained by a set of linear differential equations. Although the rate of accumulation of M intermediate is as forty times smaller than that of the native bacteriorhodopsin, the accumulation was achieved, in spite of absence of peptide bond between peptides.

1. Introduction

Purple membrane is the specific membrane found in *Halobacterium salinarium* and related species [1]. This membrane contains trimeric transmembrane protein bacteriorhodopsin *bR* in a two dimensional crystallographical fashion. Protein *bR* utilizes the energy of light to build up a proton electrochemical potential for ATP synthesis [2-4]. So a lot of effort has been made to clear up the molecular mechanism of its unique function. Several ultrafast absorption kinetic experiments revealed the primary events of the photoreaction of *bR*, that through several excited states *bR* changes from a light-adapted state to *K* intermediate within five pico seconds [5-7] at room temperature. Regarding the reaction after the primary event, it was proposed recently that *K*

IOP Conf. Series: Journal of Physics: Conf. Series **924** (2017) 012016 doi:10.1088/1742-6596/924/1/012016

intermediate decays and changes into the quasi equilibrium state between *L* intermediate and *M* intermediate within seven micro seconds, and takes an additional 130 micro seconds to change entirely into *M* intermediate [8]. And recent research on reverse photo transition *K* intermediate to *bR* [18], was reported. In previous report, it was shown that a set of reaction rates can explain the time evolution of photoreaction from *bR* to *M* intermediate [9]. By the way, protein *bR* can be reconstituted from three partial independent peptides that contain helices C-G, A, and B, respectively,[10] or that helices A-E, F, and G, respectively,[20] in spite of the absence of peptide bond [10]. Although the first type of reconstituted *b*R (*recbR*), containing helices C-G, A, and B, is known that A-B peptide interaction does not influence on the stability and photo reaction[19,20], it yields thermally, in lower temperature condition, a reversible different structural state (P_{470}) that has the absorption maximum at 470 nm, as is shown in the figure 1. A unique M_{365} intermediate yielded from P_{470} can transit directly to bR560 by light irradiation. This phenomenon is noticeable one of the important property of the first type of recbR containing helices C-G, A, and B, and the reason why this type is selected as the object of this study. However, it has not been investigated enough on the accumulation process of its *M* intermediate. In this study, under irradiation of red light, time evolution of photoreaction from native-like state of *recbR* that have the maximum wavelength of absorption at 560 nm to its *M* intermediate M_{405} was firstly and intensively analyzed.

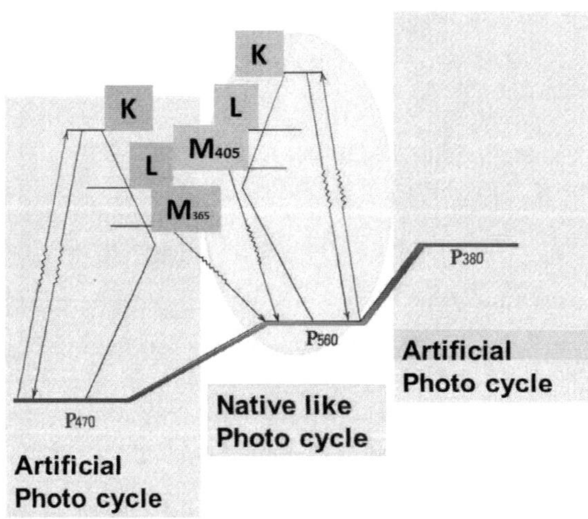

Fig 1 Energy diagram of recbR. The recbR shows three kinds of thermally stabilized structures, P_{470}, P_{560}, and P_{380} that were identified by absorption maxima .

2. Experimental section

2.1. Materials and Method

2.1.1. Preparation of peptide of bacteriorhodopsin contain helices C-G, peptides of alpha-helix A and peptides of alpha-helix B

The purple membrane fragments were purified by sucrose step gradient methods reported by Becher and Cassim [11]. Then the peptides of the helices C-G of *bR* was prepared by method of Popot [12]. The peptides of alpha-helix A and B of *bR* which were chemically synthesized, respectively were assembled with the helices C-G of *bR* in a dialysis process as is shown in the figure 2.

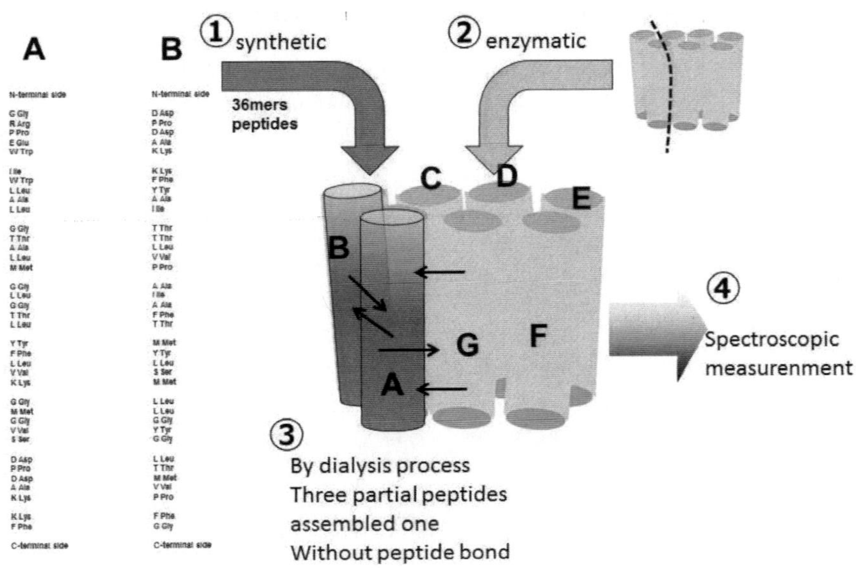

Fig 2 Procedure of preparation of recbR. The recbR. Two synthetic peptide provide helix A and B. An enzymatically cleavage peptide prpvides helices C-G, respectively. After assembling them, spectroscopic measurement was conducted at the temperature 196 K.

2.1.2. Low temperature spectroscopy

The sample for low temperature spectroscopy[13-15] contains 75% glycerol to prevent micro-crystallization of ice during cooling. For this preparation, the protein sample was pelleted at 35000 g for 60 min. The volume of the resultant protein pellet was measured with a pipette, and 3 times the volume of water of glycerol was added and used. The absorbance of about 560 nm was in the range of 0.1-0.8. A glass container was specially constructed to fit in a UV1200 spectrophotometer (Shimadzu, Kyoto). The spectrophotometer was controlled by a personal computer (Think Centre, IBM). A liquid acetone and dried ice (CO_2) was used as a coolant for maintaining the temperature at 196K. Temperature was monitored with a copper-constantan thermocouple. A 500W Xe lamp (Ushio, Tokyo) was used as an actinic light source. The wavelength of the actinic light was selected with interference and/or cut-off filters (Toshiba, Tokyo).

The sample was sealed in an optical cell composed of two quartz plates (1mm thickness) separated by a rubber o-ring (1 mm thickness). The cell was inserted into a copper cell holder, which was directly attached to the bottom of the container. The sample was illuminated with intense light ($\lambda \geqq$600nm) for at least 30 min just before cooling in order to insure that it was fully light-adapted.

2.2. Characterization

2.2.1. Reaction scheme

Primary photo-reactions of *recbR* sequentially is estimated to occur like as that of native *bR*, and can be represented by the following equations:

$$recbR_{560} + \mathrm{h}\nu \rightarrow recbR^{*}(trans) \qquad (1)$$

$$recbR^{*}(trans) \rightarrow recbR^{*}(cis) \qquad (2)$$

$$recbR(\mathrm{cis}\} \rightarrow J \qquad (3)$$

$$J \rightarrow K \qquad (4)$$

where *J*, *K* are intermediates and the indices indicate the absorption maxima of the corresponding forms and the asterisk denotes exited states. These events also occur reversely in photo equilibrium. So it is simplified as follows in terms of only *recbR* and *K*:

$$recbR_{560} \rightarrow K \qquad (5)$$

Although this state transition in room temperature is usually known as irreversible one as above, one recent study[18] reported the reverse photo transition occurs. So it is also described in the photo equilibrium state as following.

$$recbR_{560} \Leftrightarrow K$$

In the second stage, K changes to intermediate M_{405} thermally.

$$K \rightarrow L \rightarrow M_{405} \qquad (6)$$

These two stages can be summarized as follows:

$$bR_{560} \rightarrow K \rightarrow M_{405} \qquad (7)$$

At room temperature after the above reactions, the protein reverses back to bR_{560} through intermediates O and N, as follows:

$$M_{405} \rightarrow N \rightarrow O \rightarrow bR_{560} \qquad (8)$$

At the temperature around 196 K, reactions described by (8) occur at a very small rate, and the intermediate M_{405} can be accumulated. To understand the time development of M intermediate accumulation, three reaction rates u_1, u_2, u_3 can be defined for following the reactions respectively:

$$(recbR, recbR, recbR) \xrightarrow{u_1} (recbR, recbR, M)$$

$$(recbR, recbR, M) \xrightarrow{u_2} (recbR, M, M)$$

$$(recbR, M, M) \xrightarrow{u_3} (M, M, M)$$

where a bracket ( ) indicates one trimeric unit. Four kinds of trimer units (*recbR*, *recbR recbR*), (*recbR*, *recbR M*), (*recbR*, *M M*), (*M*, *M M*), can be expressed as M_0, M_1, M_2, M_3 respectively.

IOP Conf. Series: Journal of Physics: Conf. Series **924** (2017) 012016 doi:10.1088/1742-6596/924/1/012016

2.2.2. Differential equations

In order to analyze the reaction rates of conversion from bR_{560} to M_{412} a set of linear differential equations was divided as written in (eq.9-12).

$$\frac{dM_0(t)}{dt} = -u_1 M_0(t) \tag{9}$$
$$\frac{dM_1(t)}{dt} = u_1 M_0(t) - u_2 M_1(t) \tag{10}$$
$$\frac{dM_2(t)}{dt} = u_2 M_1(t) - u_3 M_2(t) \tag{11}$$
$$\frac{dM_3(t)}{dt} = u_3 M_2(t) \tag{12}$$

Initially, M_0, M_1, M_2, M_3 can be represented as follows respectively.

$$M_0(0) = 1, \quad M_1(0) = 0, \quad M_2(0) = 0, \quad M_3(0) = 0 \tag{13}$$

Under above condition, the relative number of M_{412} with irradiation time t can be defined as $M(t)$, and be represented as follows:

$$\begin{aligned} M(t) &= M_1 + 2M_2 + 3M_3 \\ &= (1 - e^{-u_1 t}) - \frac{2}{3}\left[\frac{u_1}{(u_2 - u_1)}(e^{-u_1 t} - e^{-u_2 t})\right] \\ &\quad -\frac{1}{3}\left[\frac{u_1 u_2}{(u_3 - u_1)(u_2 - u_1)}(e^{-u_1 t} - e^{-u_3 t}) + \frac{u_1 u_2}{(u_3 - u_2)(u_2 - u_1)}(e^{-u_2 t} - e^{-u_3 t})\right] \end{aligned} \tag{14}$$

IOP Conf. Series: Journal of Physics: Conf. Series **924** (2017) 012016 doi:10.1088/1742-6596/924/1/012016

3. Results and discussion

3.1. Dependency of formation of M intermediate with total irradiation time

Temperature was set at about 77 K. The intensity of light for irradiation was 7.56×10^{12} photon/sec mm^2, which does not induced a two photon-reaction during $bR \rightarrow M$. The wavelength of light for irradiation was set greater than 600 nm, which only bR absorbs. Samples were irradiated for periods of 1 min, 1 min, 2 min, 4 min, 8 min, 16 min, 32 min, 64 min, 128 min, 256 min, 512 min, 1024 min, respectively, until totally 2048 min and the absorption spectrum was measured. The absorption spectra of *recbR* are shown in Fig. 3.

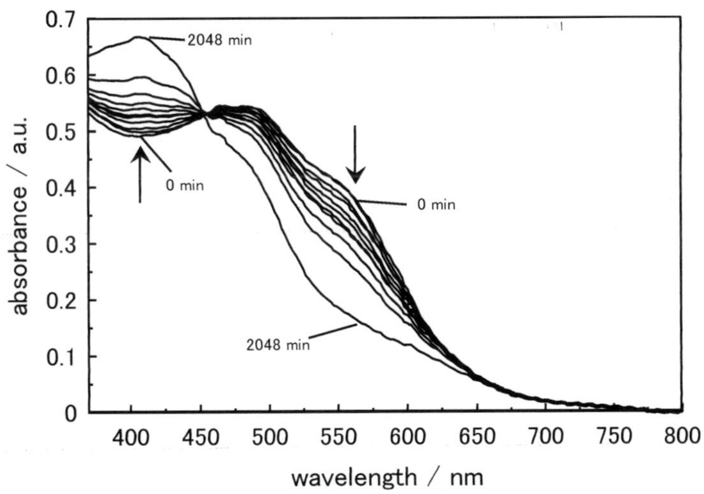

Fig 3. Absorption spectrum change of *recbR*. The absorbance of the *M* intermediate at 405 nm increased as the red light irradiation increased. On the contrary, the absorbance of the ground state of *recbR* at 560 nm decreased. At the total irradiation time by 2048 min, Almost amount of *recbR* changed to *M* intermediate.

Under irradiation a the *M* intermediate having a λ max at around 405 nm was generated. After the irradiation (total irradiation time was 2048 min), the spectrum closed to baseline. So the relative difference of absorbance at around 405 nm was normalized to 100% compared to the initial state before irradiation.

3.2. Comparison of formation rate of M intermediate of recbR with native bR

On the other hand the reaction rate of formation of *M* intermediate was simulated computationally by the equations[9]. The accumulation process of *recbR* was quietly different from native *bR*. Firstly, the formation rate is as 100 times as slower than that of native *bR*. This indicates because of the absence of peptide bond assembled protein from three peptides changes step by step to *M* intermediate. Secondly, multi-regression

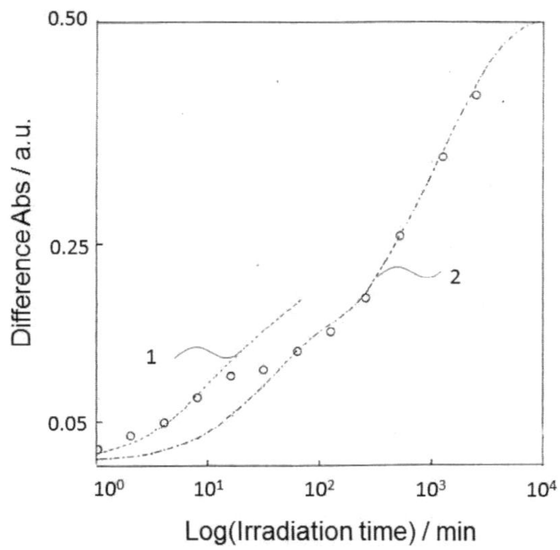

Fig. 4 The difference absorbance change at 405 nm. This value is corresponds to accumulation of *M* intermediate (o). ---: calculated data, 1 is early state change, 2 is late one.

analysis method in previous report for *nativebR* [9] could not successfully be applied on all range of time development. Then we applied the equation (14) onto an early time range and a late time range separately, and propose a model for this phenomenon to elucidate it. This is a model that there are two different modes of accumulation change. From the early state change (indicate 1 in the figure), *recbR* has changed with parameter (u1, u2, u3) = (0.04, 0.0012, 0.0012). The late state change (indicate 2 in the figure), *recbR* has changed with

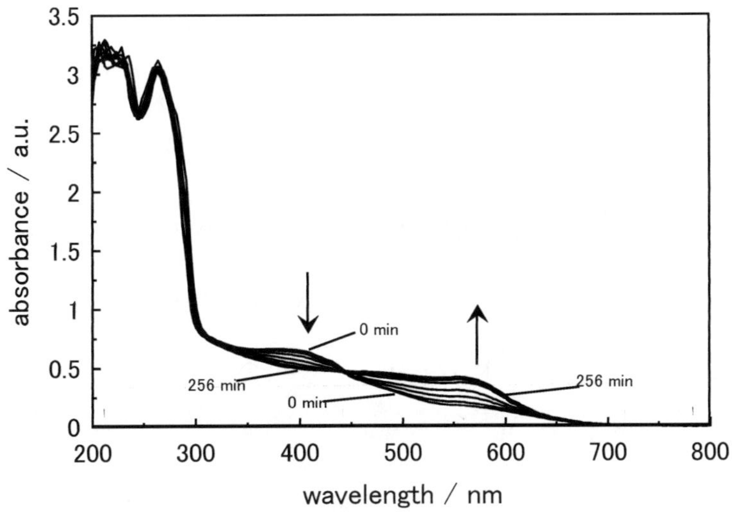

Fig.5 The absorption spectrum change of *recbR* under irradiation with 390 nm light. The absorbance changes at 405 nm, 560 nm correspond to decrease of *M* intermediate and recovery of *recbR*, respectively.

parameter (u1, u2, u3) = (0.015, 0.00045, 0.00045) that are 0.375 times as slower than early state change. Probability of total change mode is constant 1, as following.

$$M(t) = c_1 M_1(t) + c_2 M_2(t)\,, \qquad c_1 + c_1 = 1$$

These data was compared with the case of native*bR* [9]. The result of comparison was that the native*bR* changes to *M* intermediate, as forty times as faster than *recbR*.

3.3. Time development of recovery from M intermediate to ground state bR_{560} under irradiation with 390 nm light

The recovery change took 256 min. Although it was faster than that of formation of *recbR,* it indicates that small amount of energy is need for recovery from M intermediate to ground state.

3.4. Association of Change of Rate Constants, the Hypothesis of Energy Transfer in Purple Membrane and Transmembrane Helices

The reason why early state change and late state change occur is considered to be the helix-helix interaction from *recbR* to *M* intermediate transition in inter protein trimeric system. This consideration is also supported by Renthal et al.'s study that reported in it, in independent two transmembrane helices A and B, a förster resonance energy transfer occur between amino acid tryptophan in helix B and tyrosine in helix A. And from our previous morphological study of *bR* in solid film [16,17], helix-helix interaction has an essential role in the bR molecular structural change. Therefore, *recbR* constituted from peptides is, in late stage, also influenced to be delay from inter molecular interaction between helices C-G, A, and B, in inter protein trimeric system.

4. Conclusion

In this study, time development of absorbance at 560 nm and 405 nm of reconstituted bacteriorhodopsin (rec*bR*) containing helices C-G, A, and B, was analyzed for the purpose of determining reaction rates of photo reaction of *recbR* and its product *M* intermediate. The result was that *recbR* changes as forty times as slower than native *bR* in spite of absence of peptide bond.

IOP Conf. Series: Journal of Physics: Conf. Series **924** (2017) 012016 doi:10.1088/1742-6596/924/1/012016

Acknowledgements

This work was partly supported by a Grant-in Aid from the Ministry of Education, Science and Culture of Japan.

References

1. Blaurock A E.and Stoeckenius W 1971 *Nature New Biol.* **233** 152
2. Racker E and Stoeckenius W 1974 *J. Biol. Chem.* **249** 662
3. Hwang S B and Stoeckenius W 1977 *J. Mem. Biol.* **33** 325
4. Ryrie I J *et al* 1976 *Arc. Biochem. Biophys.* **176** 127
5. Nuss M C *et al* 1985 *Chem. Phys. Lett.* **117** 1
6. Sharkov A V *et al* 1985 *Biochem. Biophys. Acta.* **808** 94
7. Pollard W T *et al* 1989 J. Chem. Phys. **90** 199
8. Chizhov I *et al* 1996 *Biophys. J.* **71** 2329
9. Tsujiuchi Y *et al* 2016 *Nanosci. and Nanotech.* **16** 3431
10. Kataoka M *et al* 1992 *Photochem. and Photobiol.* **56** 895
11. Becher B and Cassim J Y 1975 *Prep. Biochem.* **5** 161
12. Popot J *et al* 1990 *Biochem.* **198** 655
13. Tokunaga F *et al* 1976 *FEBS Lett.* **72** 33
14. Iwasa T *et al* 1980 *FEBS Lett.* **101** 121
15. Iwasa T *et al* 1981 *Photochem. Photobiol.* **33** 539
16. Tsujiuchi Y *et al* 2009 *Thin Solid Films* **518** 600
17. Tsujiuchi Y and Makino Y 2011 *Physics Procedia* **14** 115
18. Feldman T B et al 2016 Photochem. Photobiol. 164 296
19. Kahn T et al 1992 Biochem. 164 296

20. Ozawa S et al 1997 BBA. 1323 145

21. Renthal R et al 2011 Biophys. chem. 159 321

IOP Conf. Series: Journal of Physics: Conf. Series **924** (2017) 012017 doi:10.1088/1742-6596/924/1/012017

Vapor Deposition Polymerization and Molecular Orientation of Polyimide Thin Films

Takatoshi Yamazaki[1], Chanya Mahapun[1], Satoshi Usui[2], Kuniaki Tanaka[1] and Hiroaki Usui[1]

[1] Graduate School of Engineering, Tokyo University of Agriculture and Technology, 2-24-16 Naka-cho, Koganei, Tokyo 184-8588, Japan

[2] Department of Environmental Science, Niigata University, 8050 Ikarashi 2-no-cho, Nishi-ku, Niigata 950-2181, Japan

E-mail: h_usui@cc.tuat.ac.jp

Abstract. Polyimide thin films were prepared by vapor deposition polymerization. Co-deposition of pyromellitic dianhydride and 1,12-diaminododecane spontaneously yielded a thin film of polyamic acid, which was imidized by annealing in nitrogen atmosphere at 200°C. It was observed that the polyimide thin film has preferential orientation to the (001) crystal plane with the molecular chain standing on the substrate. The orientation was enhanced by matching the ratio of monomer supply for vapor deposition. The molecular orientation was also influenced by modifying the substrate surface with self-assembled monolayers (SAMs). A SAM having anhydrous carboxy terminal was effective in enhancing the molecular orientation.

1. Introduction

Polyimide is a class of polymer material that has outstanding characteristics such as mechanical robustness and thermal stability. It is widely used as electrical insulator, and is an indispensable material for constructing flexible electronic circuits. In these days, polyimide insulating layers are introduced even inside the integrated circuits. Another important application of polyimide is the orientation layer for liquid crystals. For this application, the molecular orientation of polyimide is controlled by the rubbing method.

In spite of the unique properties, application of polyimide is not straightforward owing to its low processability that comes from its poor solubility and non-melting characteristics. Polyimide thin films are generally prepared by coating polyamic acid that can be dissolved in organic solvents, and then imidizing by annealing. However, residue of solvent in the film frequently causes problems. A method to overcome this limitation is the vapor-deposition polymerization, which produces polyimide thin films in high-vacuum environment without using a solvent [1, 2]. The vapor-deposition technique can alleviate the issues that originate in solvents, and is especially useful for such applications that require highly-controlled film deposition with low impurity. It is also reported that the vapor-deposition is capable of controlling the molecular orientation [3, 4]. In this paper, we report the film formation of polyimide by the vapor-deposition polymerization, and explore the molecular orientation in terms of the deposition condition. We also investigated a possibility of controlling the molecular orientation by using self-assembled monolayers (SAMs) that have reactive terminal units.

2. Experimental procedure

2.1. Vapor-deposition polymerization

Figure 1 shows the reaction scheme to form polyimide in this work. Monomers of 1,12-diaminododecane (DADD) and pyromellitic dianhydride (PMDA) were co-deposited from individual Knudsen cells in a high-vacuum chamber to form a thin film of polyamic acid. DADD was evaporated at a temperature of 80°C, and PMDA was evaporated at a temperature between 180 and 210°C. The deposition rate of each monomer was monitored by a quartz crystal microbalance, and the evaporation temperature of PMDA was adjusted to obtain a desired ratio of monomer supply. The films were deposited on glass slides coated with aluminum by vapor deposition. The substrate temperature was kept at 20°C.

The co-deposited film was exposed to the air, and annealed in nitrogen atmosphere at 100 to 250°C for 30 to 60 min to imidize by dehydration of polyamic acid. Infrared (IR) spectra of the film was measured by reflection-absorption method (HORIBA FT-730). The surface morphology was observed with an atomic force microscope (AFM: Keyence VN-8010). The crystallographic property was investigated with an X-ray diffractometer (Rigaku SmartLab, with Cu target) using the parallel beam configuration.

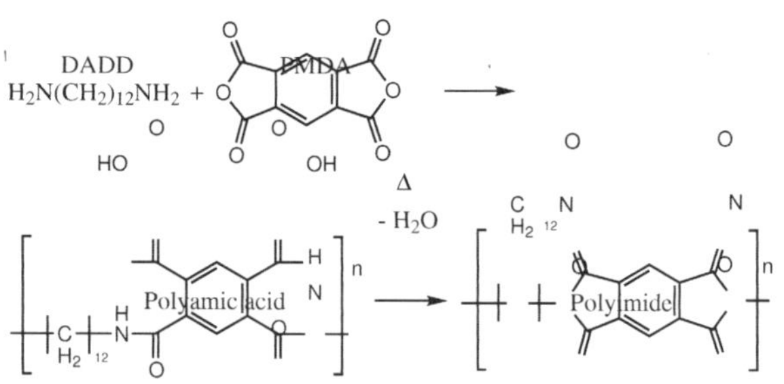

Figure 1. Reaction scheme for polyimide formation.

2.2. SAM-modification of the substrate

It is known that the surface characteristics of the substrate gives a significant influence on the film growth of vapor-deposited films. As a consequence, the molecular orientation [5, 6] and electrical characteristics [7], as well as film morphology [8] of vapor-deposited organic thin films can be controlled by modifying the substrate surface with a SAM. It is interpreted that the modification of the surface energy is a primary role of the SAM [9]. On the other hand, the authors have investigated vapor deposition of polymers on the surface of SAMs that have reactive terminal units. The reactive SAMs can induce growth of polymer brush that is covalently tethered to the inorganic surface [10, 11]. It was revealed that the covalent tethering at the interface has remarkable effect in improving the morphology and stability of the vapor-deposited films [12, 13]. This technique was also effective for controlling the film/substrate interface of organic devices [14, 15]

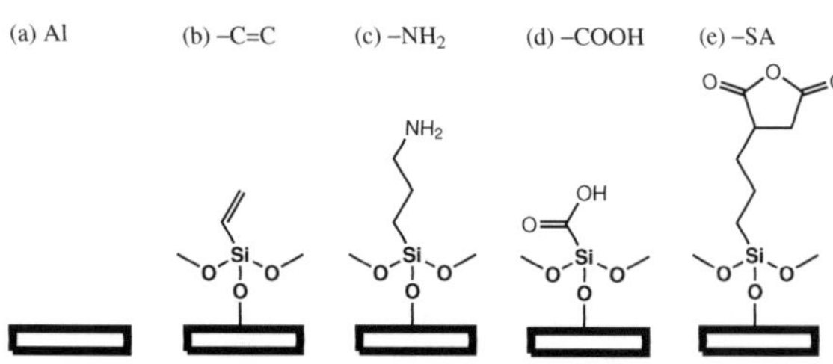

Figure 2. Structural formulae of SAMs used in this work.

IOP Conf. Series: Journal of Physics: Conf. Series **924** (2017) 012017 doi:10.1088/1742-6596/924/1/012017

In this work, four types of SAMs shown in Fig. 2 were prepared on the substrate surface to explore their influence on the molecular orientation of the vapor-deposited polyimide. The SAMs were prepared on aluminum surface that was oxidized by treating in a UV-ozone cleaner (Filgen UV253). The SAM having vinyl terminal (–C=C SAM) was prepared by immersing the substrate in 5 mM toluene solution of vinyltrimethoxysilane for 12 h. The SAM having amino terminal (–NH_2 SAM) was prepared by immersing the substrate in 1 wt% toluene solution of 3-aminopropyl trimethoxy silane for 1 h. The SAM having carboxy terminal (–COOH SAM) was prepared by oxidizing the –C=C SAM in an aqueous solution containing 0.5 mM $KMnO_4$, 19.5 mM $NaIO_4$, and 1.8 mM K_2CO_3 for 24 h. The SAM having anhydrous carboxy terminal (–SA SAM) was prepared by immersing the substrate in 3 mM toluene solution of 3-(trimethoxysilyl)propyl succinic anhydride for 3 h. All these solution processes were achieved at room temperature. After the immersing processes, the SAMs were finished by ultrasonic cleaning in toluene followed by annealing at 110°C for 1 h in nitrogen atmosphere. The SAM-modified surfaces were characterized by water contact angle measurement (First Ten Ångstrom FTA-188). On these surfaces, vapor-deposition of polyimide was achieved in the aforementioned procedure.

3. Results and discussion

3.1. Formation of polyimide thin films

A film of about 300 nm thick was obtained by co-depositing DADD and PMDA for 20 min. Figure 3 shows IR spectra of PMDA monomer (a), DADD monomer (b), and a film obtained by co-depositing PMDA and DADD at an equimolar deposition rate (c). PMDA monomer is characterized by its terminal group of cyclic acid anhydride that gives C=O stretching bands in the region of 1750 to 1880 cm^{-1} and C-O-C stretching band at 1246 cm^{-1}. DADD monomer has characteristic absorption by N-H stretching bands of amino terminal in the range of 3340 to 3150 cm^{-1}. These absorption bands of monomers were not observed in the co-deposited film. Instead, the co-deposited film showed C=O stretching band at 1643 cm^{-1} and the combination band of C-N stretching and N-H bending at 1581 cm^{-1} that are both attributed to amide bond. This result indicates that a film of polyamic acid was produced by the co-deposition of PMDA and DADD as shown by the first step of Fig. 1.

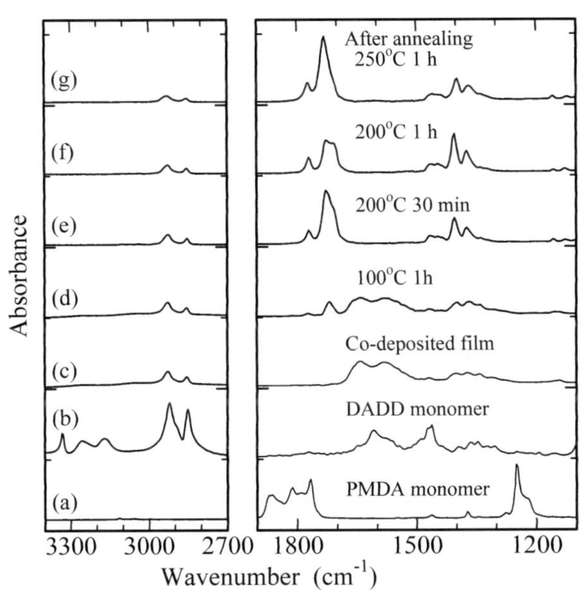

Figure 3. IR spectra of PMDA (a) and DADD (b) monomers, co-deposited film (c), and films annealed at 100°C for 1 h (d), additionally 200°C for 30 min (e), 200°C for 1 h (f), and 250°C for 1h (g).

IOP Conf. Series: Journal of Physics: Conf. Series **924** (2017) 012017 doi:10.1088/1742-6596/924/1/012017

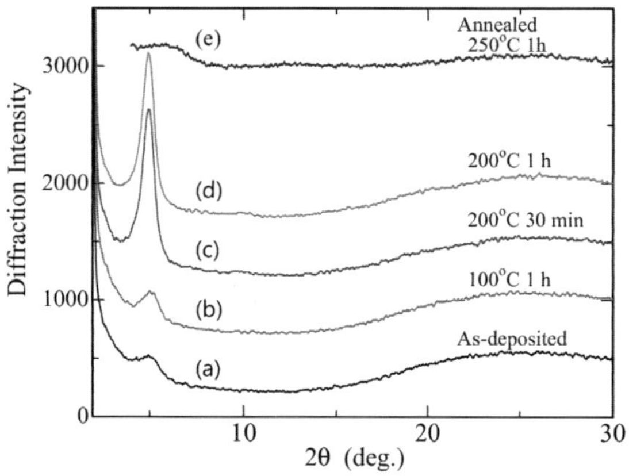

Figure 4. X-ray diffraction patterns of co-deposited film (a), and films annealed at 100°C for 1 h (b), additionally 200°C for 30 min (c), 200°C for 1 h (d), and 250°C for 1 h (e).

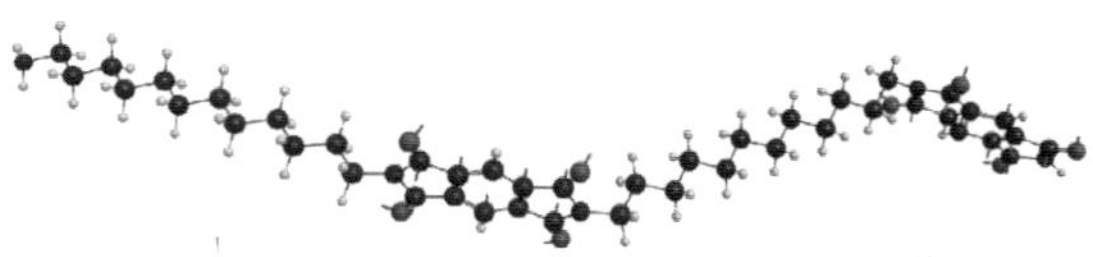

Figure 5. Molecular structure of a part of PMDA-ODA polyimide optimized by quantum chemical calculation.

As the next step, the film of polyamic acid was annealed in nitrogen atmosphere to form polyimide. Figure 3 (d) shows the IR spectra after annealing the co-deposited film for 1 h at 100°C. After this heat treatment, a new peak appeared at 1724 cm^{-1}. The IR spectra underwent further change after additional annealing at 200°C for 30 min as shown in Fig. 3 (e). The new bands at 1770 cm^{-1} and 1730 cm^{-1} can be assigned to symmetric and asymmetric stretching modes of imide C=O, and the band at 1404 cm^{-1} can be assigned as C-N stretching of cyclic imide. On the other hand, the absorption bands that originate in the amide bond disappeared after annealing at 200°C. This result indicates the formation of polyimide by cyclodehydration of polyamic acid as shown by the second step of Fig. 1. Further annealing at 200°C for 1 h did not result in significant change in IR spectra. An almost identical spectrum was obtained when the annealing was achieved at 250°C for 1 h. After the imidization, the film thickness was reduced to about 200 nm.

Figure 4 shows X-ray diffraction patterns of the co-deposited film (a) and the film after annealing under the different conditions (b–e). After the imidization, a prominent peak appeared at $2\theta = 5.0°$, which corresponds to a plane distance of 1.77 nm. This diffraction represents the (001) plane of monoclinic polyimide having alkyl chain in its backbone [16, 17]. The result of X-ray diffraction analysis indicates that the polyimide thin film prepared by the vapor-deposition polymerization has preferential crystal orientation with the molecular chain standing on the substrate surface. Figure 5 shows an optimized structure of the two repeating units of PMDA-DADD polyimide obtained by the semi empirical quantum chemical calculation (MOPAC PM3). This model gave the unit length of the PMDA-DADD imide as 2.21 nm. It suggests that the polymer backbone of the vapor-deposited polyimide is inclined by 37° from the substrate normal. When the annealing was performed at 250°C, the film lost its preferential orientation as shown by pattern (e) in Fig. 4. It is likely that the annealing temperature of 250°C was higher than the glass transition temperature of this polymer [18]. Therefore, the optimum temperature for imidization in this system is considered to be 200°C.

3.2. Effect of monomer supply ratio

The polymerization of polyimide proceeds by stepwise reaction of polyaddition process. With this type of reaction, the degree of polymerization is known to be largely influenced by the ratio

IOP Conf. Series: Journal of Physics: Conf. Series **924** (2017) 012017 doi:10.1088/1742-6596/924/1/012017

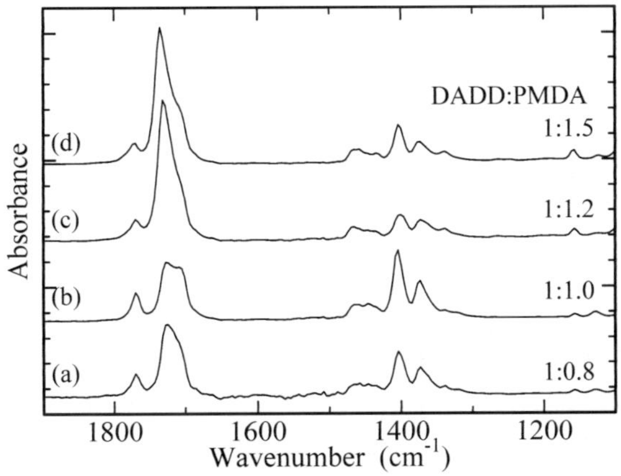

Figure 6. IR Spectra of polyimide films deposited with DADD:PMDA monomer supply ratio of 1:0.8 (a), 1:1 (b), 1:1.2 (c), and 1:1.5 (d).

Table 1. Influence of ratio of monomer supply to the ratio of IR absorbance of symmetric/asymmetric vibrations of C=O. Arithmetic average surface roughness R_a of the polyimide film is also shown in the last line.

DADD:PMDA	0.8	1.0	1.2	1.5
symmetric/assymmetric	0.36	0.52	0.20	0.21
R_a (nm)	0.4	0.6	0.9	1.0

of monomer supply. Figure 6 shows IR spectra of polyimide films prepared with molar supply ratio of DADD:PMDA of 1:0.8 (a), 1:1 (b), 1:1.2 (c), and 1:1.5 (d). The films were annealed at 100°C for 1 h followed by 200°C for 1 h. All the films co-deposited in this range of supply ratio produced polyimide without appreciable residue of monomers or polyamic acid. DADD and PMDA monomers have low sticking coefficients by themselves, and surplus monomers tend to reevaporate spontaneously from the substrate surface. In other words, only those molecules that reacted to form the polymer accumulated stably on the substrate surface. For this reason, polyimide films were obtained without being crucially influenced by the monomer supply ratio.

On the other hand, a detailed inspection of Fig. 6 reveals that the monomer supply ratio influences the relative absorbance of the symmetric and asymmetric stretching bands of imide C=O at 1770 and 1730 cm^{-1}. It should be noted that the reflection absorption measurement employed in this work has preferential sensitivity to the transition dipole moment in the direction normal to the substrate surface, which means that the absorbance is influenced by the molecular orientation. The symmetric stretching mode of imide C=O has its transition dipole moment in parallel with the polymer backbone, while the transition dipole moment of the asymmetric stretching is in perpendicular to the backbone. Table 1 shows the ratio of symmetric/asymmetric absorption bands of imide C=O for the films deposited with different monomer supply ratio. The film deposited with equimolar monomer supply showed the most dominant absorption of the symmetric band, indicating that this film has the highest orientation with the molecular chain standing on the substrate surface.

The effect of monomer supply ratio was observed more explicitly in the X-ray diffraction patterns as shown in Fig. 7. The film deposited with equimolar monomer supply showed the most prominent orientation to the (001) plane. When the supply ratio deviated from the equimolar condition, the diffraction intensity of (001) plane was significantly reduced, while a new peak appeared at $2\theta = 20°$, which corresponds to either (011) or (101) plane of the monoclinic polyimide crystal. This result is in good accordance with the observation by the IR spectra. With

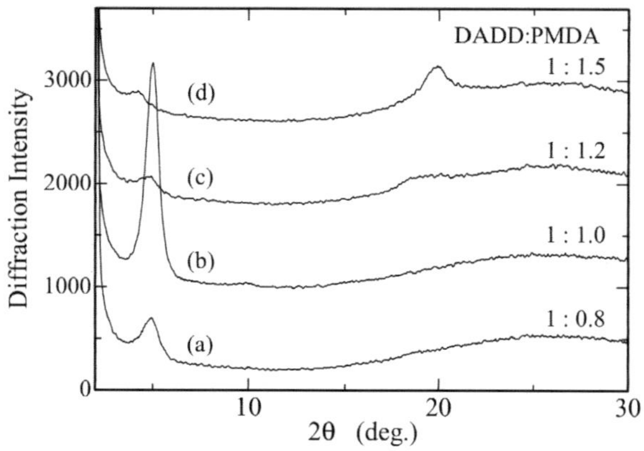

Figure 7. X-ray diffraction patterns of polyimide films deposited with DADD:PMDA monomer supply ratios of 1:0.8 (a), 1:1 (b), 1:1.2 (c), and 1:1.5 (d).

the stepwise polymerization, it is generally known that the degree of polymerization increases by properly matching the monomer supply. On the other hand, it is also known that vapor deposited rod-like molecules have a tendency to orient with their molecular axes standing on the substrate surface [19]. Therefore, it is considered that the proper balance of monomer supply results in polyimide of a longer chain, which has higher tendency to orient in normal direction.

Surface morphology of the polyimide films deposited with different ratio of monomer supply is shown in Fig. 8. The arithmetic average surface roughness R_a of these films is listed in the last line of Table 1. The vapor-deposited polyimide films have excellent smoothness. However, R_a increased slightly when there was surplus supply of PMDA. This is because PMDA monomer is less volatile than DADD, and has a higher possibility of affecting the film morphology when supplied in surplus amount.

3.3. Effect of SAM modification on the molecular orientation

The film growth morphology by the vapor deposition is primarily governed by the balance of surface energies and interface energy of the film and substrate materials. That is the main reason why SAMs are frequently employed to control the growth of organic thin films on inorganic substrates [20]. On the other hand, the authors have shown that covalent tethering at the film/substrate interface is an effective strategy to control the interfacial characteristics of organic

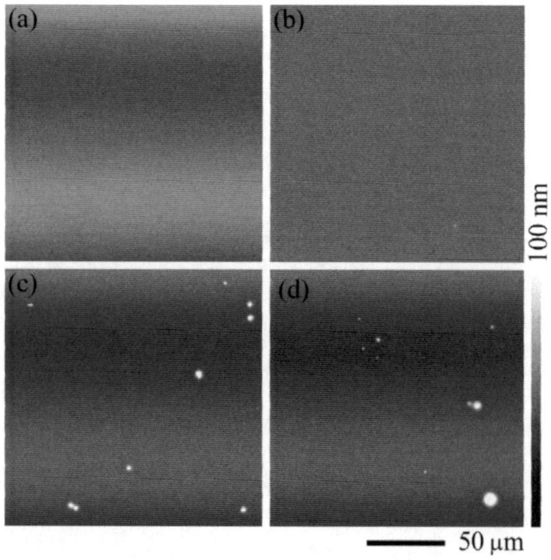

Figure 8. AFM images of polyimide films deposited with DADD:PMDA monomer supply ratios of 1:0.8 (a), 1:1 (b), 1:1.2 (c), and 1:1.5 (d).

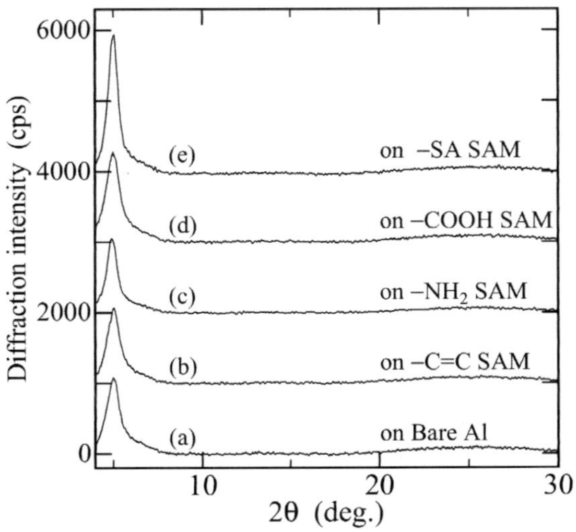

Figure 9. X-ray diffraction patterns of polyimide thin films deposited on bare aluminum (a), and on aluminum modified with SAMs having terminal groups of –C=C (b), –NH_2 (c), –COOH (d), and –SA (e).

Table 2. Water contact angles and surface roughness of bare and SAM-modified aluminum surfaces. Surface roughness of polyimide deposited on these surfaces are also listed in the last line.

Substrate	Al	–C=C SAM	–NH_2 SAM	–COOH SAM	–SA SAM
Contact angle (deg)	72.8	76.4	51.4	55.5	53.4
R_a of substrate (nm)	0.5	1.0	0.9	0.9	1.3
R_a of polyimide (nm)	0.5	0.9	0.6	0.7	1.1

thin films. This was achieved by using SAMs that have reactive terminal units [21]. From this standpoint, the substrate surface was modified with different SAMs shown in Fig. 2, and the vapor-deposition polymerization was achieved on their surfaces.

Figure 9 shows X-ray diffraction patterns of polyimide thin films deposited on a bare aluminum (a) and on the aluminum modified with SAMs having –C=C (b), –NH_2 (c), –COOH (d), and –SA (e) terminal groups. All the films showed preferential crystal orientation to (001) plane as described in previous section. Above all, the film deposited on –SA SAM showed the strongest diffraction peak. It is frequently pointed out that the SAMs influences the film growth through modification of surface energy. Table 2 shows the water contact angles of the bare and the SAM-modified aluminum surfaces. The clean aluminum was hydrophobic and so was the surface modified with –C=C SAM. The reactive SAMs having –NH_2, –COOH, and –SA terminals were less hydrophobic than the bare aluminum, but there was no significant difference in water contact angles between these reactive SAMs. Therefore, the pronounced molecular orientation observed specifically on the –SA SAM cannot be explained only by the factor of surface energy.

From a chemical viewpoint, –SA group has higher reactivity than the –COOH group. The difference in reactivity can be an important factor to influence the molecular orientation. With a purpose to compare the chemical reactions of the SAMs, quantum chemical calculations were performed by the MOPAC (PM3) program for the reactions of –NH_2 SAM with PMDA (a), –COOH SAM with DADD (b), and –SA SAM with DADD (c). The optimized structures of these reaction products are shown in Fig. 10. The enthalpies for these reactions were calculated to be -52.0 kJ/mol for –NH_2 with PMDA, +32.3 kJ/mol for –COOH with DADD, and -34.3

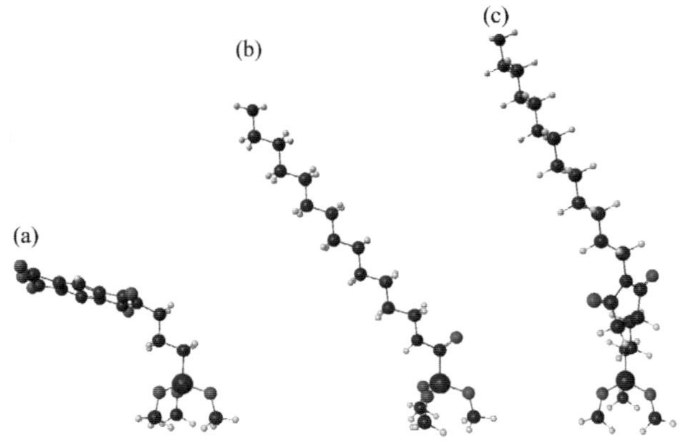

Figure 10. Optimized structures by quantum chemical calculations for $-NH_2$ SAM reacted with PMDA (a), –COOH SAM reacted with DADD (b), and –SA SAM reacted with DADD (c).

kJ/mol for –SA with DADD. It is considered that the –COOH SAM is the least convenient of these three SAMs for causing the reaction. In general, the reaction of a carboxylic acid with an amine requires a dehydration reagent, although the vacuum environment is dehydrating itself. As far as the enthalpy change is concerned, the $-NH_2$ SAM appears to be the best of these three. However, the optimized structure shown in Fig. 10(a) implies that the reaction product of $-NH_2$ SAM and PMDA has a bent structure, which is not convenient for enhancing perpendicular molecular orientation of the polyimide chain. Therefore, it is considered that the –SA SAM is appropriate both for tethering the interface and for enhancing the molecular orientation.

Figure 11 shows the surface morphology of the polyimide films deposited on the different surfaces. All the films had smooth surfaces. The surface roughness R_a of these films are listed

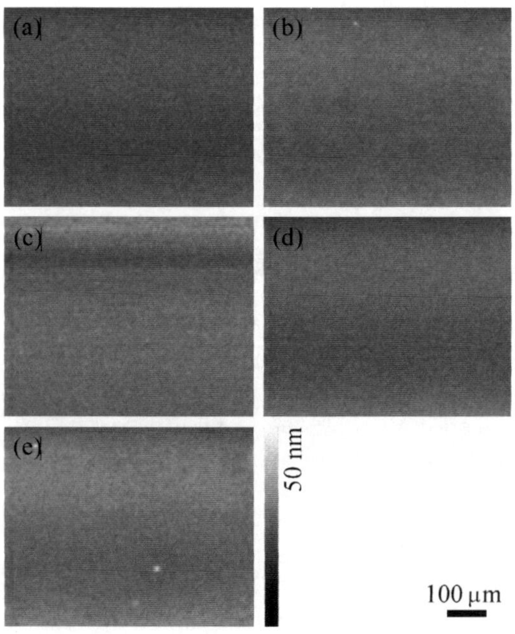

Figure 11. AFM images of polyimide films deposited on bare aluminum (a), and on those modified with SAMs of –C=C (b), $-NH_2$ (c), –COOH (d), and –SA (e).

in Table 2 together with the R_a of the SAM-modified substrates. The film on –SA SAM seems to have slightly larger surface roughness, but it might originate in the roughness of the SAM itself. There was essentially no significant difference in film morphology by the SAM modification.

4. Conclusions

Polyimide thin films were obtained by the vapor-deposition polymerization method. Co-deposition of PMDA and DADD monomers yielded a film of polyamic acid in the as-deposited state, which was imidized by annealing at 200°C in nitrogen atmosphere. The polyimide film had preferential crystal orientation with the (001) plane parallel to the substrate surface, which means that the polymer backbone is standing on the substrate. The ratio of monomer supply did not give a crucial influence on the reaction of polyimide formation. On the other hand, the molecular orientation was enhanced by matching the ratio of monomer supply to equimolar condition. It was also found that the SAM modification of the substrate surface is effective for controlling the molecular orientation. Above all, the SAM having an anhydrous carboxy terminal was effective in enhancing the molecular orientation. The major role of this SAM is considered to originate in the chemical reaction at the film/substrate interface rather than the simple difference in surface energy.

Acknowledgments

This research was partially supported by a Grant-in-Aid for Scientific Research (No. 25286022) from the Japan Society for the Promotion of Science.

References

[1] Salem J R, Sequeda F O, Duran J, Lee W Y and Yang R M 1986 *J. Vac. Sci. Technol. A* **4** 369–374
[2] Takahashi Y, Iijima M, Inagawa K and Itoh A 1987 *J. Vac. Sci. Technol. A* **5** 2253–2256
[3] Kubono A, Higuchi H, Umemoto S and Okui N 1993 *Thin Solid Films* **229** 133–136
[4] Kubono A, Okui N, Tanaka K, Umemoto S and Sakai T 1991 *Thin Solid Films* **199** 385–393
[5] Kim D H, Lee H S, Yang H, Yang L and Cho K 2008 *Adv. Funct. Mater.* **18** 1363–1370
[6] Virkar A, Mannsfeld S, Oh J H, Toney M F, Tan Y H, Liu G, Scott J C, Miller R and Bao Z 2009 *Adv. Funct. Mater.* **19** 1962–1970
[7] Gundlach D J, Nichols J A, Zhou L and Jackson T N 2002 *Appl. Phys. Lett.* **80** 2925–2927
[8] Shtein M, Mapel J, Benziger J B and Forrest S R 2002 *Appl. Phys. Lett.* **81** 268–270
[9] Yang S Y, Shin K and Park C E 2005 *Adv. Funct. Mater.* **15** 1806–1814
[10] Katsuki K, Bekku H, Kawakami A, Locklin J, Patton D, Tanaka K, Advincula R and Usui H 2005 *Jpn. J. Appl. Phys.* **44** 504–508
[11] Duran H, Ogura K, Nakao K, Vianna S D B, Usui H, Advincula R C and Knoll W 2009 *Langmuir* **25** 10711–10718
[12] Umemoto Y, Kim S H, Advincula R C, Tanaka K and Usui H 2010 *Jpn. J. Appl. Phys.* **49** 04DK21/1–5
[13] Hagihara Y, Kim S H, Tanaka K, Advincula R C and Usui H 2014 *Jpn. J. Appl. Phys.* **53** 04EK04/1–5
[14] Kawakami A, Katsuki K, Advincula R C, Tanaka K, Ogino K and Usui H 2008 *Jpn. J. Appl. Phys.* **47** 3156–3161
[15] Ono S, Usui S, Kim S H, Tanaka K, Advincula R C and Usui H 2016 *J. Nanosci. Nanotechnol.* **16** 3407–3413
[16] Kazaryan L G, Azriel A E and Vasil'ev V A 1979 *Vysokomolekulyarnye Soedineniya B* **21** 644–645
[17] Kazaryan L G, Azriel A E, Vasil'ev V A, Annenkova N G and Pinaeva N K 1988 *Vysokomolekulyarnye Soedineniya A* **30** 644–648
[18] Dong L, Chiu Y C, Chueh C C, Yu A D and Chen W C 2014 *Polym. Chem.* **5** 6834–6846
[19] Kubono A and Akiyama R 2002 *Mol. Cryst. Liq. Cryst. A* **378** 167–183
[20] Turak A 2013 *Optoelectronics - Advanced Materials and Devices* ed Pyshkin S L and Ballato J M (Rijeka, Croatia: Intech) chap 10, pp 229–268
[21] Ono S, Usui S, Tanaka K and Usui H 2017 *Jpn. J. Appl. Phys.* **56** 04CL08/1–6